STRONG STABILITY PRESERVING RUNGE–KUTTA AND MULTISTEP TIME DISCRETIZATIONS

T0324824

STRONG STABILITY PRESERVING RUNGE–KUTTA AND MULTISTEP TIME DISCRETIZATIONS

Sigal Gottlieb
University of Massachusetts Dartmouth, USA

David Ketcheson
KAUST, Kingdom of Saudi Arabia

Chi-Wang Shu
Brown University, USA

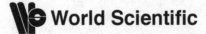 **World Scientific**

NEW JERSEY · LONDON · SINGAPORE · BEIJING · SHANGHAI · HONG KONG · TAIPEI · CHENNAI

Published by

World Scientific Publishing Co. Pte. Ltd.

5 Toh Tuck Link, Singapore 596224

USA office: 27 Warren Street, Suite 401-402, Hackensack, NJ 07601

UK office: 57 Shelton Street, Covent Garden, London WC2H 9HE

Library of Congress Cataloging-in-Publication Data
Gottlieb, Sigal.
  Strong stability preserving Runge-Kutta and multistep time discretizations / by Sigal Gottlieb,
David Ketcheson & Chi-Wang Shu.
    p. cm.
  Includes bibliographical references and index.
  ISBN-13: 978-9814289269 (hard cover : alk. paper)
  ISBN-10: 9814289264 (hard cover : alk. paper)
  1. Runge-Kutta formulas. 2. Differential equations--Numerical solutions. 3. Stability.
I. Ketcheson, David I. II. Shu, Chi-Wang. III. Title.
  QA297.5.G68 2011
  518'.6--dc22

                                                                    2010048026

British Library Cataloguing-in-Publication Data
A catalogue record for this book is available from the British Library.

Printed in Singapore.

In Memory of
DAVID GOTTLIEB
November 14, 1944 - December 6, 2008

Preface

Strong stability preserving (SSP) high order time discretizations were developed to ensure nonlinear stability properties necessary in the numerical solution of hyperbolic partial differential equations with discontinuous solutions. SSP methods preserve the strong stability properties – in any norm, seminorm or convex functional – of the spatial discretization coupled with first order Euler time stepping. Explicit strong stability preserving (SSP) Runge–Kutta methods have been employed with a wide range of spatial discretizations, including discontinuous Galerkin methods, level set methods, ENO methods, WENO methods, spectral finite volume methods, and spectral difference methods. SSP methods have proven useful in a wide variety of application areas, including (but not limited to): compressible flow, incompressible flow, viscous flow, two-phase flow, relativistic flow, cosmological hydrodynamics, magnetohydrodynamics, radiation hydrodynamics, two-species plasma flow, atmospheric transport, large-eddy simulation, Maxwell's equations, semiconductor devices, lithotripsy, geometrical optics, and Schrödinger equations. These methods have now become mainstream, and a book on the subject is timely and relevant. In this book, we present SSP time discretizations from both the theoretical and practical points of view. Those looking for an introduction to the subject will find it in Chapters 1, 2, 8, 10, and 11. Those wishing to study the development and analysis of SSP methods will find Chapters 3, 4, 5, 8, and 9 of particular interest. Finally, those looking for practical methods to use will find them in Chapters 2, 4, 6, 7, and 9.

We are very grateful to our colleague Colin Macdonald who contributed to much of the research in this book, especially in Chapters 5, 7, and 9. We also wish to thank Randy LeVeque for his guidance and for encouraging the collaboration that led to this book.

Words are not sufficient to express our thanks to our families, who have shared in the process of writing this book. Their support and encouragement made this work possible, and the joy they bring to our lives make it more meaningful.

Much of the research that led to this book was funded by the U.S. Air Force Office of Scientific Research, under grants FA9550-06-1-0255 and FA9550-09-1-0208. We wish to express our gratitude to Dr. Fariba Fahroo of the AFOSR for her support of recent research in the field of SSP methods, and for her encouragement of and enthusiasm about this book project.

This work is dedicated to the memory of David Gottlieb, whose wisdom, kindness, and integrity continue to be an inspiration to those who knew him.

Sigal Gottlieb, David Ketcheson, & Chi-Wang Shu
October 2010

Contents

ix

Chapter 1

Overview: The Development of SSP Methods

In this book we discuss a class of time discretization techniques, which are termed strong stability preserving (SSP) time discretizations and have been developed over the past 20 years as an effective tool for solving certain types of large ordinary differential equation (ODE) systems arising from spatial discretizations of partial differential equations (PDEs).

The numerical solution of ODEs is of course an established research area. There are many well-studied methods, such as Runge–Kutta methods and multistep methods. There are also many excellent books on this subject, for example [8, 64, 83]. ODE solvers for problems with special stability properties, such as stiff ODEs, are also well studied, see, e.g., [32, 17].

The SSP methods are also ODE solvers, therefore they can be analyzed by standard ODE tools. However, these methods were designed specifically for solving the ODEs coming from a spatial discretization of time-dependent PDEs, especially hyperbolic PDEs. The analysis of SSP methods can be facilitated when this background is taken into consideration.

In the one-dimensional scalar case, a hyperbolic conservation law is given by

$$u_t + f(u)_x = 0 \tag{1.1}$$

where u is a function of x and t, and the subscripts refer to partial derivatives; for example, $u_t = \frac{\partial u}{\partial t}$. Hyperbolic PDEs pose particular difficulties for numerical methods because their solutions typically contain discontinuities. We refer the readers to, e.g. the book of Smoller [94] and the lecture notes of LeVeque [68] for an introduction to hyperbolic conservation laws and their numerical solutions. We would like to find a numerical approximation, still denoted by u with an abuse of notation, which discretizes the spatial derivative (e.g. $f(u)_x$ in (1.1)) in the PDE. The result is then a

1

semi-discrete scheme, i.e. an ODE in the time variable t:

$$u_t = F(u). \tag{1.2}$$

There is typically a small parameter Δx in the spatial discretization, for example the mesh size in a finite difference, finite volume or discontinuous Galerkin finite element scheme, which goes to zero with an increased desire for accuracy. Typically, we would have

$$F(u) = -f(u)_x + O(\Delta x^k) \tag{1.3}$$

for some positive integer k, which is referred to as the spatial order of accuracy for the semi-discrete scheme (1.2) approximating the PDE (1.1). The system of ODEs obtained this way is usually very large, since the size of the system depends on the spatial discretization mesh size Δx: the smaller Δx, the larger the size of the ODE system.

We would now like to discretize the large ODE system (1.2). Stability is considered when the size of the system grows without bound with the decrease of the spatial discretization mesh size Δx. More importantly, there are certain stability properties of the original PDE, such as total variation stability or maximum norm stability, which may be maintained by the semi-discrete scheme (1.2). It is also often the case that such stability properties are maintained by the fully discrete scheme

$$u^{n+1} = u^n + \Delta t F(u^n) \tag{1.4}$$

which is a first order forward Euler approximation of the ODE (1.2). A few important examples of the stability properties used in applications will be given in Chapter 11. The problem with the fully discrete scheme (1.4) is that it is only first order accurate in time. For hyperbolic problems, the linear stability requirement usually leads to a bounded ratio between the time step Δt and the spatial mesh size Δx. This leads to a global first order accuracy for the scheme (1.4) regardless of the high order spatial accuracy (1.3). It is therefore highly desirable to increase the order of accuracy in time, while still maintaining the same stability properties. SSP methods are designed to achieve this goal.

We can describe the main property of SSP time discretizations in this way: if we *assume* that the first order, forward Euler time discretization (1.4) of a method of lines semi-discrete scheme (1.2) is stable under a certain norm (or a semi-norm or a convex functional)

$$\|u^{n+1}\| \le \|u^n\|, \tag{1.5}$$

then a SSP high order time discretization *maintains* this stability under a suitable restriction on the time step.

One might ask the question whether it is worthwhile and necessary to use SSP time discretizations. We will argue in favor of the necessity or advantage of SSP methods through a numerical example, originally given in [26].

Example 1.1. We consider the nonlinear Burgers equation, namely (1.1) with $f(u) = \frac{u^2}{2}$, with Riemann initial data:

$$u(x,0) = \begin{cases} 1, & \text{if } x \leq 0 \\ -0.5, & \text{if } x > 0. \end{cases} \tag{1.6}$$

We use a second order minmod based MUSCL spatial discretization [106] to approximate the spatial derivative in (1.1). It is easy to prove, by using Harten's lemma [34] (see Chapter 11), that the forward Euler time discretization with this second order MUSCL spatial operator is total variation diminishing (TVD) under the Courant-Friedrichs-Levy (CFL) condition:

$$\Delta t \leq \frac{\Delta x}{2 \max_j |u_j^n|}. \tag{1.7}$$

Thus $\Delta t = \frac{\Delta x}{2 \max_j |u_j^n|}$ is used in the calculation.

We consider two second order Runge–Kutta methods for the time discretization. The first is the second order SSP Runge–Kutta method given in [92]:

$$u^{(1)} = u^n + \Delta t F(u^n) \tag{1.8}$$
$$u^{n+1} = \frac{1}{2}u^n + \frac{1}{2}u^{(1)} + \frac{1}{2}\Delta t F(u^{(1)}),$$

the second is the method:

$$u^{(1)} = u^n - 20\Delta t F(u^n) \tag{1.9}$$
$$u^{n+1} = u^n + \frac{41}{40}\Delta t F(u^n) - \frac{1}{40}\Delta t F(u^{(1)}).$$

It is easy to verify that both methods are second order accurate in time. However, it can be easily verified that the second method (1.9) is not SSP. In Figure 1.1 (reproduced from [92]) we show the results of the SSP Runge–Kutta method (1.8) and the non-SSP method (1.9), after the shock moves about 50 grids. We can clearly see that the non-SSP result is oscillatory (there is an overshoot).

Such oscillations are also observed when the non-SSP Runge–Kutta method coupled with a second order TVD MUSCL spatial discretization is

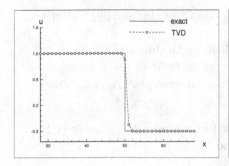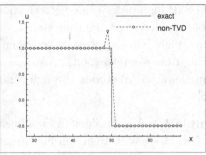

Fig. 1.1   Second order TVD MUSCL spatial discretization. Solution after 500 time steps. Left: SSP time discretization (1.8); Right: non-SSP time discretization (1.9).

applied to a linear PDE $(u_t + u_x = 0)$. Moreover, for some Runge–Kutta methods, if one looks at the intermediate stages, e.g. $u^{(1)}$ (1.8) or (1.9), one observes even bigger oscillations. Such oscillations may render difficulties when physical problems are solved, such as the appearance of negative density and pressure for Euler equations of gas dynamics. On the other hand, SSP Runge–Kutta methods guarantee that each intermediate stage solution is also TVD.

Another example demonstrating the advantage of SSP methods versus non-SSP methods for shock wave calculations can be found in [31]. Further examples comparing some commonly used non-SSP methods with SSP methods may be found in, e.g., [62, 55, 54]. These numerical examples demonstrate that it is at least safer to use SSP time discretizations whenever possible, especially when solving hyperbolic PDEs with shocks, which make the traditional linear stability analysis inadequate. In terms of computational cost, most SSP methods are of the same form and have the same cost as traditional ODE solvers. It is true that the time step Δt might need to be smaller to prove the SSP property than, say, when linear stability is proven, however in many situations Δt can be taken larger in practical calculations without causing instability.

Development of SSP methods was historically motivated in two ways, and developed by two groups: one focusing on hyperbolic partial differential equations, the other focusing on ordinary differential equations. Many terms have been used to describe what we refer to as strong stability preservation; here we stick mostly to this term for clarity.

SSP time discretization methods were first developed by Shu in [91] and by Shu and Osher in [92], and were called TVD time discretizations. The terminology was adopted because the method of lines ODE (1.2) and its forward Euler time-discretized version (1.4) both satisfy the total variation diminishing property when applied to scalar one dimensional nonlinear hyperbolic conservation laws (1.1). Shu and Osher also proposed the idea of using downwind-biased discretizations in order to preserve strong stability; see Chapter 10. In [92], a class of second to fifth order Runge–Kutta time discretizations are proven to be SSP. In [91], a class of high order multistep SSP methods are given, as well as a class of first order SSP Runge–Kutta methods that have very large SSP coefficients. Later, Gottlieb and Shu [26] performed a systematic study of Runge–Kutta SSP methods, proving the optimality of the two-stage, second order and three-stage, third order SSP Runge–Kutta methods as well as finding low storage three-stage, third order SSP Runge–Kutta methods, and proving that four-stage, fourth order SSP Runge–Kutta methods with non-negative coefficients cannot exist. In [27], Gottlieb, Shu and Tadmor reviewed and further developed SSP Runge–Kutta and multistep methods. It was in this paper that the term "strong stability preserving", or SSP, was first used. In later literature the terms SSP time discretizations and TVD time discretizations have been used simultaneously. The new results in [27] include the optimal explicit SSP linear Runge–Kutta methods, their application to the strong stability of coercive approximations, a systematic study of explicit SSP multistep methods, and the study of the strong stability preserving property of implicit Runge–Kutta and multistep methods. Spiteri and Ruuth [97] found a new class of SSP Runge–Kutta methods with larger SSP coefficient by allowing the number of stages to be greater than the order of accuracy. The same authors also proved that luck runs out in this approach starting from fifth order: there is no SSP fifth order Runge–Kutta method with non-negative coefficients [86]. Gottlieb and Gottlieb in [28] obtained optimal linear SSP Runge–Kutta methods when the number of stages is larger than the order of accuracy. They have also made an interesting application of these methods to certain special variable coefficient ODEs, such as those coming from spatial discretizations of linear, constant coefficient PDEs (e.g. Maxwell's equations) with time dependent boundary conditions. Ruuth and Spiteri used the Shu-Osher theory and numerical optimization to develop optimal methods over many classes, including downwind methods [97, 87, 84]. In [71], Liu, Shu and Zhang studied SSP properties for the deferred correction methods, which can be considered as a special class of Runge–Kutta methods.

Among the ODE community, related work began with investigations of positivity by Bolley and Crouzeix [5] and of contractivity (or monotonicity) by Spijker [95], for linear systems of ODEs. In these works it was noted that such properties cannot be preserved unconditionally by Runge–Kutta or multistep methods of higher than first order. Conditional strong stability preservation was shown to be related to the radius of absolute monotonicity for methods satisfying a circle condition. Optimal Runge–Kutta methods for linear systems, including implicit and explicit methods, were investigated in [60, 104]. Conditions for strong stability preserving linear multistep methods in the context of nonlinear equations were given in [89], and optimal linear multistep methods for linear and nonlinear equations were investigated by Lenferink [66, 67]. The rich theory of absolute monotonicity of Runge–Kutta methods, and its relation to contractivity for nonlinear equations was developed by Kraaijevanger [62]. In addition, Kraaijevanger's work provided important results such as the order barriers for SSP Runge–Kutta methods and several optimal methods. The relation of this theory to positivity preservation was later developed by Horvath [40, 41].

The equivalence of the Shu-Osher theory and the theory of absolute monotonicity, both of which had been well developed for about 15 years, was discovered independently and almost simultaneously by Ferracina and Spijker [19, 21] and by Higueras [35, 37]. The unification of the two theories has provided a theoretical framework that is more elegant, complete, and useful than either of its predecessors.

Recently, SSP theory has been extended in several important ways. Higueras has extended the theory of absolute monotonicity to include methods with downwind-biased operators [37] and, more generally, additive Runge–Kutta methods [38, 39]. A theory of SSP has been developed also for diagonally split Runge–Kutta methods, which lie outside the class of general linear methods and are capable of being unconditionally SSP and higher than first order accurate [3, 43, 4, 40, 73]. Hundsdorfer et al. have developed a class of linear multistep methods that satisfy a more general (weaker) condition than the SSP condition, but allow much larger timesteps [49, 48, 85]. First attempts to characterize the practical sharpness of SSP theory have been made in [53, 31]. New approaches to finding optimal methods have yielded new optimal methods in several classes, which in many cases are accompanied by memory-efficient implementations, see [22, 55, 54, 56, 57]. Gottlieb and Ruuth developed methods aimed at taking advantage of efficiently-implemented downwind discretizations [30]. A general SSP theory for multi-stage methods applied to nonlinear equations has been developed by Spijker [96], and optimal SSP general linear methods have been investigated for certain classes in [46, 15, 58].

SSP time discretizations are widely used in numerical solutions of time dependent PDEs, especially hyperbolic PDEs. Essentially non-oscillatory (ENO) and weighted ENO (WENO) finite difference and finite volume schemes in, e.g. [92, 93, 79, 80, 51, 44, 2], and Runge–Kutta discontinuous Galerkin finite element methods in, e.g. [13, 12, 14] are examples. Other examples of applications include the total variation bounded spectral methods [9] and the weighted L^2 SSP higher order discretizations of spectral methods [25]. In fact, the (semi) norm can be replaced by any convex function, as the arguments of SSP are based on convex decompositions of high order methods in terms of the first order Euler method. An example of this is the cell entropy stability property of high order schemes studied in [81] and [75].

The SSP property is a very strong requirement that guarantees strong stability (monotonicity) in arbitrary convex functionals, for arbitrary starting values and arbitrary nonlinear, non-autonomous equations, as long as the forward Euler method satisfies the desired monotonicity condition. The result of this strong stability requirement is a rather stringent restriction on the time step and, in some cases, barriers on the allowable order of a SSP method. However, in many cases the SSP condition is too strong. For example, when dealing with smooth, well-resolved problems, weaker conditions may guarantee positivity [41, 42]. Also, if instead of considering arbitrary convex functionals we require monotonicity in some inner-product norm, we can obtain larger time step restrictions and, in the implicit case, methods which break the order barrier [36]. In the case of multistep methods, clever choices of starting methods may relax the time step restriction (Chapter 8). Finally, as we see in Chapter 4, if we require strong stability preservation only when integrating linear autonomous equations (i.e. a linear SSP property), the time step condition is also more relaxed. When none of these simplifications apply, as is the case for nonlinear PDEs with discontinuous solution, we turn to the SSP property to guarantee strong stability in the desired norm.

In the following chapters we will present the details of the algorithm development and analysis of SSP methods. This book is organized as follows. Chapters 2-7 focus on SSP properties of Runge–Kutta methods. In Chapter 2 we describe the context in which SSP methods were first developed and provide the basic understanding of explicit SSP Runge–Kutta methods as convex combinations of forward Euler steps. In Chapter 3 we study the SSP coefficient, which is a measurement of the time step restriction, hence the efficiency, of the SSP methods. Chapter 4 is devoted to the study of SSP methods for linear problems with constant coefficients. In Chapter 5, we collect some observations about bounds on the SSP coefficient and barriers on the order of methods with positive SSP coefficients. In Chapter 6, we study a special class of SSP methods which require low storage, hence are suitable for large scale high-dimensional calculations. Chapter 7 contains the discussion on implicit SSP Runge–Kutta methods. Multistep methods are discussed in Chapter 8, and two-step Runge–Kutta methods in Chapter 9. In Chapter 10, we study a special class of SSP methods, which have negative coefficients and so require a modification of the spatial discretization to achieve stability. Finally, sample applications of the SSP methods to particular stability properties are given in Chapter 11.

Chapter 2

Strong Stability Preserving Explicit Runge–Kutta Methods

2.1 Overview

In this chapter we describe the context in which SSP methods were first developed and provide a basic understanding of explicit SSP Runge–Kutta methods as convex combinations of forward Euler steps. We explain the connections between a representation of Runge–Kutta methods first introduced in [92], which facilitates the analysis of SSP properties and is later referred to as the Shu-Osher form, and then proceed to present some optimal SSP explicit Runge–Kutta methods.

2.2 Motivation

As mentioned in the previous chapter, the concept of strong stability preserving time stepping methods first arose in the numerical solution of hyperbolic partial differential equations with discontinuous solutions. When we wish to approximate the solution of a hyperbolic conservation law such as

$$u_t + f(u)_x = 0, \tag{2.1}$$

we use the method-of-lines approach. We first use some spatial discretization, denoted $-F(u)$, to approximate the spatial derivative $f(u)_x$. This yields a semi-discrete system of ordinary differential equations

$$u_t = F(u), \tag{2.2}$$

where, with some abuse of notation, u in the scheme (2.2) is a vector of approximations to the exact solution $u(x, t)$ of the PDE (2.1), for example in a finite difference approximation, the jth element $u_j(t)$ in (2.2) is an approximation to the point value $u(x_j, t)$ in (2.1). Now we apply some time stepping method to obtain the solution of the system of ordinary differential equations, and we use the notation u^n to denote the solution to the fully discrete system at time t^n, for example in a finite difference approximation, $u_j^n \approx u(x_j, t^n)$. The critical questions are: for which combinations of spatial and time discretizations is this process stable and convergent? What analysis is appropriate? What if the solution is discontinuous – does our numerical solution still make sense?

Let us look at a simple example: the linear one-way wave equation.

Example 2.1.

$$u_t + u_x = 0 \qquad (2.3)$$

with initial condition $u(x, 0) = g(x)$ and a periodic boundary condition. For this equation, the exact solution is simply the initial condition convected around the domain, $u(x, t) = g(x - t)$. To approximate the solution numerically, we can discretize the spatial derivative by a first order backward difference

$$u_x = \frac{u_j - u_{j-1}}{\Delta x} \qquad (2.4)$$

and the resulting system of ordinary differential equation can be solved using a time-integration scheme. If we use forward Euler's method, for example, we get

$$u_j^{n+1} = u_j^n - \frac{\Delta t}{\Delta x} \left(u_j^n - u_{j-1}^n \right). \qquad (2.5)$$

A standard von Neumann L^2 stability analysis of this method shows that it is stable for $\frac{\Delta t}{\Delta x} \leq 1$. This is a linear problem, and if the solution is smooth (i.e. if the initial condition is smooth), L^2 linear stability (given by the von Neumann stability analysis) is necessary and sufficient for convergence ([100] Theorem 1.5.1).   □

This analysis makes sense in this case because the PDE is linear. But what if the PDE is nonlinear? In that case, if a numerical method is consistent and its *linearization* is L^2 stable and adequately dissipative, then for sufficiently *smooth* problems the nonlinear approximation is convergent [99].

But if the PDE is nonlinear and the solution is *not* smooth, linear stability analysis is not enough for convergence. Furthermore, L^2 stability does not tell the whole story even in the linear case – the numerical solution may be oscillatory but not unstable in the linear case. For hyperbolic partial differential equations with discontinuous solutions, L^2 linear stability analysis is not enough. For these equations, we need something more powerful: a theory which will guarantee nonlinear stability in the maximum norm or in the bounded variation (or total variation) semi-norm, or at least some non-oscillatory property. However, the problem is that unlike linear stability, which can often be studied directly even for complex time discretizations, nonlinear stability is more difficult to examine. For this reason, a tremendous amount of effort has been put into the development of high order spatial discretizations $F(u)$ that, *when coupled with the forward Euler time stepping method*, have the desired nonlinear stability properties for approximating discontinuous solutions of hyperbolic partial differential equations (see, e.g. [34, 77, 101, 13, 63, 102, 70]). However, for actual computation, higher order time discretizations are usually needed and there is no guarantee that the spatial discretization will still possess the same nonlinear stability property when coupled with a linearly stable higher order time discretization.

So our challenge is to find higher order time discretizations that do indeed guarantee that these nonlinear stability properties satisfied by the spatial discretization when coupled with forward Euler integration will be *preserved* when the same spatial discretization is coupled with these higher order methods.

To summarize: the idea behind strong stability preserving methods is to begin with a method-of-lines semi-discretization that is strongly stable in a certain norm, semi-norm, or convex functional under forward Euler time stepping, when the time step Δt is suitably restricted, and then try to find a higher order time discretization that maintains strong stability for the same norm, perhaps under a different time step restriction. In other words, given a semi-discretization of the form (2.2) and convex functional $\| \cdot \|$, we assume that there exists a value Δt_{FE} such that

$$\|u + \Delta t F(u)\| \le \|u\| \text{ for } 0 \le \Delta t \le \Delta t_{FE}, \text{ for all } u. \tag{2.6}$$

We say that the method is *strong stability preserving* (SSP) with SSP coefficient $\mathcal{C}$ if (in the solution of (2.2)) it holds that

$$\|u^{n+1}\| \le \|u^n\|, \tag{2.7}$$

whenever (2.6) holds and the time step satisfies

$$\Delta t \le \mathcal{C} \Delta t_{FE}. \tag{2.8}$$

2.3　SSP methods as convex combinations of Euler's method: the Shu-Osher formulation

Explicit SSP Runge–Kutta methods were first introduced by Shu and Osher in [92] to guarantee that spatial discretizations that are total variation diminishing (TVD) and total variation bounded (TVB) when coupled with forward Euler will still produce TVD and TVB solutions when coupled with these higher order Runge–Kutta methods. It is for this reason that SSP methods were initially known as TVD methods. The key observation in the development of SSP methods was that certain Runge–Kutta methods can be written as convex combinations of Euler's method, so that any convex functional properties of Euler's method will carry over to these Runge–Kutta methods.

An explicit Runge–Kutta method is commonly written in the Butcher form

$$u^{(i)} = u^n + \Delta t \sum_{j=1}^{m} a_{ij} F(u^{(j)}) \quad (1 \le i \le m) \tag{2.9}$$

$$u^{n+1} = u^n + \Delta t \sum_{j=1}^{m} b_j F(u^{(j)}).$$

It can also be written in the form

$$u^{(0)} = u^n,$$

$$u^{(i)} = \sum_{j=0}^{i-1} \left(\alpha_{ij} u^{(j)} + \Delta t \beta_{ij} F(u^{(j)}) \right), \qquad 1 \le i \le m \tag{2.10}$$

$$u^{n+1} = u^{(m)},$$

where consistency requires that $\sum_{j=0}^{i-1} \alpha_{ij} = 1$. This form (often called the Shu-Osher form) is convenient because, if all the coefficients α_{ij} and β_{ij} are non-negative, it can easily be manipulated into convex combinations of forward Euler steps, with a modified time step. This observation motivates the following theorem:

Theorem 2.1. *([92] Section 2). If the forward Euler method applied to* (2.2) *is strongly stable under the time step restriction* $\Delta t \le \Delta t_{\mathrm{FE}}$, *i.e.* (2.6) *holds, and if* $\alpha_{ij}, \beta_{ij} \ge 0$, *then the solution obtained by the Runge–Kutta method* (2.10) *satisfies the strong stability bound*

$$\|u^{n+1}\| \le \|u^n\|$$

under the time step restriction

$$\Delta t \leq \mathcal{C}(\boldsymbol{\alpha}, \boldsymbol{\beta})\Delta t_{FE}, \tag{2.11}$$

where $\mathcal{C}(\boldsymbol{\alpha}, \boldsymbol{\beta}) = \min_{i,j} \frac{\alpha_{ij}}{\beta_{ij}}$, *and the ratio is understood as infinite if* $\beta_{ij} = 0$.

Proof. Each stage of the Runge–Kutta method (2.10) can be rewritten as a convex combination of forward Euler steps. Since all α_{ij} are non-negative and $\sum_{j=0}^{i-1} \alpha_{ij} = 1$, we have by convexity of $\| \cdot \|$ that

$$\|u^{(i)}\| = \left\| \sum_{j=0}^{i-1} \left(\alpha_{ij} u^{(j)} + \Delta t \beta_{ij} F(u^{(j)}) \right) \right\|$$

$$= \left\| \sum_{j=0}^{i-1} \alpha_{ij} \left(u^{(j)} + \Delta t \frac{\beta_{ij}}{\alpha_{ij}} F(u^{(j)}) \right) \right\|$$

$$\leq \sum_{j=0}^{i-1} \alpha_{ij} \left\| u^{(j)} + \Delta t \frac{\beta_{ij}}{\alpha_{ij}} F(u^{(j)}) \right\|.$$

Now, since each $\|u^{(j)} + \Delta t \frac{\beta_{ij}}{\alpha_{ij}} F(u^{(j)})\| \leq \|u^{(j)}\|$ as long as $\frac{\beta_{ij}}{\alpha_{ij}} \Delta t \leq \Delta t_{FE}$, and using again that $\sum_{j=0}^{i-1} \alpha_{ij} = 1$ by consistency, we have $\|u^{(i)}\| \leq \|u^n\|$ for each stage as long as $\frac{\beta_{ij}}{\alpha_{ij}} \Delta t \leq \Delta t_{FE}$ for all i and j. In particular, this yields $\|u^{n+1}\| \leq \|u^n\|$. ∎

Remark 2.1. We note that the proof breaks down if any of the coefficients are negative. However, if the α_{ij} coefficients are non-negative and only the β_{ij} coefficients can become negative, it is possible to construct SSP methods using convex combinations of forward Euler and downwinded (or backward-in-time) Euler.

Given the ODEs $u_t = F(u)$ and $u_t = \tilde{F}(u)$, where F and $\tilde{F}$ are spatial discretizations that both approximate the term $-f(u)_x$ in (2.1), satisfying $\|u^n + \Delta t F(u^n)\| \leq \|u^n\|$ and $\|u^n - \Delta t \tilde{F}(u^n)\| \leq \|u^n\|$ respectively, under the time step restriction $\Delta t \leq \Delta t_{\text{FE}}$, and if $\alpha_{ij} \geq 0$, then the solution obtained by the Runge–Kutta method (2.10) where $F(u^{(j)})$ is replaced by $\tilde{F}(u^{(j)})$ whenever the corresponding β_{ij} is negative, satisfies the strong stability bound

$$\|u^{n+1}\| \leq \|u^n\|$$

under the time step restriction

$$\Delta t \leq \mathcal{C}(\boldsymbol{\alpha}, \boldsymbol{\beta})\Delta t_{FE}, \tag{2.12}$$

where $\mathcal{C}(\boldsymbol{\alpha}, \boldsymbol{\beta}) = \min_{i,k} \frac{\alpha_{ij}}{|\beta_{ij}|}$, and the ratio is understood as infinite if $\beta_{ij} = 0$. The proof here follows as above, but using the fact that if β_{ij} is negative, that step is a downwinded Euler step. Methods which feature the downwinded operators will be presented in Chapter 10. Until that chapter, we consider only SSP methods with non-negative coefficients α_{ij}, β_{ij}. □

It is interesting that the stable time step is the product of only two factors, the forward Euler time step Δt_{FE}, which depends on the spatial discretization alone, and the coefficient $\mathcal{C}(\boldsymbol{\alpha}, \boldsymbol{\beta})$, which depends only on the time discretization. In the literature, $\mathcal{C}$ has been referred to as a *CFL coefficient*. However, we avoid this usage because the CFL condition prescribes a relation between the time step and the spatial grid size, whereas the SSP coefficient describes the ratio of the strong stability preserving time step to the strongly stable forward Euler time step.

Theorem 2.1 gives a sufficient time step restriction for the solution to satisfy the strong stability bound (2.7), but does not address whether this condition is necessary. Furthermore, it does not tell us how to go about finding SSP Runge–Kutta methods, nor how to find the method which has the largest *SSP coefficient* $\mathcal{C}$ possible. This is further complicated by the fact that several representations of the same method can lead us to different conclusions about the SSP coefficient of a method.

Example 2.2. Consider the two-stage second order Runge–Kutta method, based on the trapezoidal rule:

$$u^{(0)} = u^n \tag{2.13a}$$

$$u^{(1)} = u^{(0)} + \Delta t F(u^{(0)}) \tag{2.13b}$$

$$u^{n+1} = u^{(0)} + \frac{1}{2} \Delta t F(u^{(0)}) + \frac{1}{2} \Delta t F(u^{(1)}). \tag{2.13c}$$

Note that this is in the Shu-Osher form (2.10) and yields $\mathcal{C}(\boldsymbol{\alpha}, \boldsymbol{\beta}) = 0$. However, using the equation for $u^{(1)}$, we can rewrite the equation for u^{n+1} to obtain a better result. For instance,

$$u^{n+1} = \frac{3}{4} u^{(0)} + \frac{1}{4} \Delta t F(u^{(0)}) + \frac{1}{4} u^{(1)} + \frac{1}{2} \Delta t F(u^{(1)}) \tag{2.14}$$

yields $\mathcal{C}(\boldsymbol{\alpha}, \boldsymbol{\beta}) = 1/2$. This is still not optimal; rewriting (2.13c) as

$$u^{n+1} = \frac{1}{2} u^{(0)} + \frac{1}{2} u^{(1)} + \frac{1}{2} \Delta t F(u^{(1)}) \tag{2.15}$$

yields $\mathcal{C}(\boldsymbol{\alpha}, \boldsymbol{\beta}) = 1$. In fact, this is the optimal value of $\mathcal{C}(\boldsymbol{\alpha}, \boldsymbol{\beta})$ for this method. □

2.4 Some optimal SSP Runge–Kutta methods

Much research in the area of SSP methods focuses on finding methods with the largest possible SSP coefficient, so as to allow for the largest time step. The process of searching for optimal methods has significantly increased in sophistication over the last decade, aided by the discovery of connections between the SSP coefficient and the radius of absolute monotonicity. However, the first SSP methods found, namely the two-stage second order and three-stage third order Runge–Kutta methods, were found by a simple numerical search [92]. They were later proven optimal among all Runge–Kutta methods with their respective order and number of stages in [26] by algebraic arguments. In the following sections we present some optimal methods of orders two through four.

2.4.1 *A second order method*

In this section, we look at the class of two-stage second order Runge–Kutta methods, and show that the optimal SSP coefficient for all possible representations of these methods in the form (2.10) is $\mathcal{C} = 1$.

Theorem 2.2. [26]. *If we require $\beta_{ij} \geq 0$, then an optimal second order SSP Runge–Kutta method (2.10) is given by*

$$u^{(1)} = u^n + \Delta t L(u^n) \tag{2.16}$$
$$u^{n+1} = \frac{1}{2}u^n + \frac{1}{2}u^{(1)} + \frac{1}{2}\Delta t L(u^{(1)}).$$

We will refer to this method as **SSPRK(2,2)**.

Proof. All two-stage second order explicit Runge–Kutta methods can be written in the Shu-Osher form

$$u^{(1)} = u^n + \Delta t \beta_{10} L(u^n)$$
$$u^{n+1} = \alpha_{20} u^n + \alpha_{21} u^{(1)} + \beta_{20} \Delta t L(u^n) + \beta_{21} \Delta t L(u^{(1)}).$$

If we choose β_{10} and α_{21} as free parameters, the order conditions mean that the other coefficients are [92]:

$$\begin{cases} \alpha_{10} = 1 \\ \alpha_{20} = 1 - \alpha_{21} \\ \beta_{20} = 1 - \frac{1}{2\beta_{10}} - \alpha_{21}\beta_{10} \\ \beta_{21} = \frac{1}{2\beta_{10}} \end{cases} \tag{2.17}$$

Assume that we can obtain an SSP coefficient $\mathcal{C} > 1$. To have this happen, $\alpha_{10} = 1$ implies $\beta_{10} < 1$, which in turn implies $\frac{1}{2\beta_{10}} > \frac{1}{2}$. Also, $\alpha_{21} > \beta_{21} = \frac{1}{2\beta_{10}}$, which implies $\alpha_{21}\beta_{10} > \frac{1}{2}$. We would thus have

$$\beta_{20} = 1 - \frac{1}{2\beta_{10}} - \alpha_{21}\beta_{10} < 1 - \frac{1}{2} - \frac{1}{2} = 0,$$

which is a contradiction.

While this method is optimal among all second order SSP Runge–Kutta methods of two stages, there exist second order methods which have more stages and a larger SSP coefficient. These can be found in Chapters 4 and 6.

2.4.2   *A third order method*

In this section, we look at the class of three-stage third order Runge–Kutta methods, and show that the optimal SSP coefficient for all possible representations of these methods in the form (2.10) is $\mathcal{C} = 1$.

Theorem 2.3. *If we require $\beta_{ij} \geq 0$, an optimal third order SSP Runge–Kutta method (2.10) is given by*

$$u^{(1)} = u^n + \Delta t L(u^n)$$
$$u^{(2)} = \frac{3}{4}u^n + \frac{1}{4}u^{(1)} + \frac{1}{4}\Delta t L(u^{(1)})$$
$$u^{n+1} = \frac{1}{3}u^n + \frac{2}{3}u^{(2)} + \frac{2}{3}\Delta t L(u^{(2)}).$$

We refer to this method as **SSPRK(3,3)**.

Proof.   Any three-stage third order method can be written in the Shu-Osher form, or equivalently in the Butcher-like array form of

$$u^{(1)} = u^n + a_{21}L(u^n)$$
$$u^{(2)} = u^n + a_{31}\Delta t L(u^n) + a_{32}\Delta t L(u^{(1)})$$
$$u^{(n+1)} = u^n + b_1\Delta t L(u^n) + b_2\Delta t L(u^{(1)}) + b_3\Delta t L(u^{(2)}) \qquad (2.18)$$

where the relationship between the coefficients in (2.18) and the α_{ij} and β_{ij} in the Shu-Osher form (2.10) is:

$$a_{21} = \beta_{10}$$
$$a_{31} = \beta_{20} + \alpha_{21}\beta_{10}$$
$$a_{32} = \beta_{21}$$
$$b_1 = \alpha_{32}\alpha_{21}\beta_{10} + \alpha_{31}\beta_{10} + \alpha_{32}\beta_{20} + \beta_{30}$$
$$b_2 = \alpha_{32}\beta_{21} + \beta_{31} \qquad\qquad (2.19)$$
$$b_3 = \beta_{32}.$$

The third order conditions can be solved to obtain a two-parameter family as well as two special cases of one-parameter families [83]. The proof follows each case separately. The important observation is that all the α_{ij} and β_{ij} coefficients are required to be non-negative, which implies that the Butcher coefficients must also be non-negative. Furthermore, an SSP coefficient $\mathcal{C} > 1$ requires that $\alpha_{ij} > \beta_{ij} \geq 0$ unless both of them are zeroes. We also require that $a_{32} = \beta_{21} > 0$, $a_{21} = \beta_{10} > 0$, and $b_3 = \beta_{32} > 0$, because any of those being zero would mean that we have fewer stages. In the following we show that a SSP coefficient $\mathcal{C} > 1$ is not consistent with non-negative coefficients.

General Case: We define parameters α_2 and α_3, where $\alpha_3 \neq \alpha_2$, $\alpha_3 \neq 0$, $\alpha_2 \neq 0$ and $\alpha_2 \neq \frac{2}{3}$, the coefficients become

$$a_{21} = \alpha_2 \qquad\qquad b_1 = 1 + \frac{2 - 3(\alpha_2 + \alpha_3)}{6\alpha_2\alpha_3}$$

$$a_{31} = \frac{3\alpha_2\alpha_3(1-\alpha_2)-\alpha_3^2}{\alpha_2(2-3\alpha_2)} \qquad b_2 = \frac{3\alpha_3 - 2}{6\alpha_2(\alpha_3 - \alpha_2)}$$

$$a_{32} = \frac{\alpha_3(\alpha_3-\alpha_2)}{\alpha_2(2-3\alpha_2)} \qquad\qquad b_3 = \frac{2 - 3\alpha_2}{6\alpha_3(\alpha_3 - \alpha_2)}.$$

From these we see that $6\alpha_2 a_{32} b_3 = 1$ and $a_{31} + a_{32} = \alpha_3$. Now,

$$c > 1 \Rightarrow a_{21} = \beta_{10} < \alpha_{10} = 1$$

together with $a_{21} > 0$, this implies $0 < \alpha_2 < 1$.

Since $\alpha_3 \neq \alpha_2$ we separate this into two cases: In the first case we consider $\alpha_3 > \alpha_2$: $a_{32} \geq 0 \Rightarrow \alpha_2 < \frac{2}{3}$ and $b_2 \geq 0 \Rightarrow \alpha_3 \geq \frac{2}{3}$.

$\beta_{20} \geq 0$ and $\alpha_{21} > \beta_{21}$ imply $a_{31} \geq \alpha_{21}\beta_{10} > \beta_{21}\beta_{10}$, which is $\alpha_3 - a_{32} > a_{32}\alpha_2$, or $\frac{\alpha_3}{1+\alpha_2} > a_{21}$. So we must have

$$\alpha_3 < \frac{3\alpha_2 - 2\alpha_2^2}{1 + \alpha_2}.$$

On the other hand, $\beta_{31} \geq 0$ requires $b_2 \geq \alpha_{32}\beta_{21} > c_{32}c_{21} = \frac{1}{6\alpha_2}$, which is $3\alpha_3 - 2 > \alpha_3 - \alpha_2$, or

$$\alpha_3 > 1 - \frac{1}{2}\alpha_2.$$

Combining these two inequalities, we get $1 - \frac{1}{2}\alpha_2 < \frac{3\alpha_2 - 2\alpha_2^2}{1+\alpha_2}$, or $(2 - 3\alpha_2)(1-\alpha_2) < 0$, which is a contradiction, since $2 - 3\alpha_2 > 0$ and $1 - \alpha_2 > 0$.

In the second case we look at $\alpha_2 > \alpha_3$. Observe that $\alpha_3 = a_{31} + a_{32} > 0$ requires $\alpha_3 > 0$.

$b_3 > 0$ requires $\alpha_2 > \frac{2}{3}$, and $b_2 \geq 0$ requires $\alpha_3 \leq \frac{2}{3}$.

$b_2 \geq \alpha_{32}\beta_{21} > b_3 a_{32} = \frac{1}{6\alpha_2}$, which is

$$\alpha_3 < 1 - \frac{1}{2}\alpha_2.$$

$a_{31} \geq \alpha_{21}\beta_{10} > \beta_{21}\beta_{10}$ requires

$$\alpha_3 > \frac{\alpha_2(3 - 2\alpha_2)}{1 + \alpha_2}.$$

Putting these two inequalities together, we have $\frac{\alpha_2(3-2\alpha_2)}{1+\alpha_2} < 1 - \frac{1}{2}\alpha_2$, which means $(2 - 3\alpha_2)(1 - \alpha_2) > 0$, a contradiction since $1 - \alpha_2 > 0$ and $2 - 3\alpha_2 < 0$.

Special Case I: If $\alpha_2 = \alpha_3 = \frac{2}{3}$, the coefficients satisfy

$$a_{21} = \frac{2}{3} \qquad b_1 = \frac{1}{4}$$

$$a_{31} = \frac{2}{3} - \frac{1}{4\omega_3} \qquad b_2 = \frac{3}{4} - \omega_3$$

$$a_{32} = \frac{1}{4\omega_3} \qquad b_3 = \omega_3.$$

$\beta_{31} \geq 0$ and $\alpha_{32} > \beta_{32} = b_3$ requires $b_2 \geq \alpha_{32}\beta_{21} > b_3\beta_{21} = \frac{1}{4}$ which implies $\omega_3 < \frac{1}{2}$.

$\beta_{20} \geq 0$ and $\alpha_{21} > \beta_{21} = a_{21}$ requires $a_{31} \geq \alpha_{21}\beta_{10} > \frac{2}{3}a_{32}$, which means $\frac{2}{3} - \frac{1}{4\omega_3} > \frac{2}{3}\frac{1}{4\omega_3}$, for which we must have $\omega_3 > \frac{5}{8}$. A contradiction.

Special Case II: If $\alpha_3 = 0$, the coefficients satisfy

$$a_{21} = \frac{2}{3} \qquad b_1 = \frac{1}{4} - \omega_3$$

$$a_{31} = \frac{1}{4\omega_3} \qquad b_2 = \frac{3}{4}$$

$$a_{32} = -\frac{1}{4\omega_3} \qquad b_3 = \omega_3.$$

Clearly a_{31} and a_{32} cannot be simultaneously non-negative.

Special Case III: $\alpha_2 = 0$, the method is not third order. $\qquad\qquad \square$

SSPRK(3,3) is widely known as the Shu-Osher method, and is probably the most commonly used SSP Runge–Kutta method. Although this method is only third order accurate, it is most popular because of its simplicity, low-storage properties, and its classical linear stability properties.

Although both the SSPRK(2,2) and SSPRK(3,3) methods have SSP coefficient $C = 1$, which permits a time step of the same size as forward Euler would permit, it is clear that the computational cost is double and triple (respectively) that of the forward Euler. Thus, we find it useful to define the *effective SSP coefficient* as $C_{\text{eff}} = \frac{C}{m}$, where m is the number of function evaluations required per time step (typically, m is equal to the number of stages). In the case of SSPRK(2,2) and SSPRK(3,3) the effective SSP coefficient is $C_{\text{eff}} = \frac{1}{2}$ and $C_{\text{eff}} = \frac{1}{3}$, respectively. While in this case the increase in the order of the method makes this additional computational cost acceptable, the notion of the effective SSP coefficient is useful when comparing two methods of the same order that have different numbers of stages and different SSP coefficients.

For a given order of accuracy, if we allow more stages, we can obtain a larger SSP coefficient. More importantly, optimal SSPRK methods with more stages typically have larger *effective* SSP coefficient. For instance, the optimal five-stage, third order method SSPRK(5,3)

$$
\begin{aligned}
u^{(1)} &= u^n + 0.37726891511710 \Delta t F(u^n) \\
u^{(2)} &= u^{(1)} + 0.37726891511710 \Delta t F(u^{(1)}) \\
u^{(3)} &= 0.56656131914033 u^n + 0.43343868085967 u^{(2)} \\
&\quad + 0.16352294089771 \Delta t F(u^{(2)}) \\
u^{(4)} &= 0.09299483444413 u^n + 0.00002090369620 u^{(1)} \\
&\quad + 0.90698426185967 u^{(3)} + 0.00071997378654 \Delta t F(u^n) \qquad (2.20) \\
&\quad + 0.34217696850008 \Delta t F(u^{(3)}) \\
u^{(5)} &= 0.00736132260920 u^n + 0.20127980325145 u^{(1)} \\
&\quad + 0.00182955389682 u^{(2)} + 0.78952932024253 u^{(4)} \\
&\quad + 0.00277719819460 \Delta t F(u^n) + 0.00001567934613 \Delta t F(u^{(1)}) \\
&\quad + 0.29786487010104 \Delta t F(u^{(1)})
\end{aligned}
$$

has $C_{\text{eff}} \approx 0.53$, larger than that of SSPRK(3,3).

2.4.3   *A fourth order method*

Although optimal two-stage second order and three-stage third order methods were easy to find, the search for a four-stage fourth order SSP Runge–Kutta method was fruitless. In [62, 26] it was proved that all four-stage, fourth order Runge–Kutta methods with positive SSP coefficient $\mathcal{C}$ must have at least one negative β_{ij}.

We begin by writing the four-stage, fourth order Runge–Kutta method in the following form:

$$
\begin{aligned}
u^{(1)} &= u^n + c_{10}L(u^n) \\
u^{(2)} &= u^n + c_{20}\Delta t L(u^n) + c_{21}\Delta t L(u^{(1)}) \\
u^{(3)} &= u^n + c_{30}\Delta t L(u^n) + c_{31}\Delta t L(u^{(1)}) + c_{32}\Delta t L(u^{(2)}) \\
u^{n+1} &= u^n + c_{40}\Delta t L(u^n) + c_{41}\Delta t L(u^{(1)}) + c_{42}\Delta t L(u^{(2)}) + c_{43}\Delta t L(u^{(3)})
\end{aligned}
\tag{2.21}
$$

there is the following relationship between the coefficients c_{ij} here and α_{ij} and β_{ij} in the Shu-Osher form (2.10):

$$
\begin{aligned}
c_{10} &= \beta_{10} \\
c_{20} &= \beta_{20} + \alpha_{21}\beta_{10} \\
c_{21} &= \beta_{21} \\
c_{30} &= \alpha_{32}\alpha_{21}\beta_{10} + \alpha_{31}\beta_{10} + \alpha_{32}\beta_{20} + \beta_{30} \\
c_{31} &= \alpha_{32}\beta_{21} + \beta_{31} \\
c_{32} &= \beta_{32} \\
c_{40} &= \alpha_{43}\alpha_{32}\alpha_{21}\beta_{10} + \alpha_{43}\alpha_{32}\beta_{20} + \alpha_{43}\alpha_{31}\beta_{10} + \alpha_{42}\alpha_{21}\beta_{10} \\
&\quad + \alpha_{41}\beta_{10} + \alpha_{42}\beta_{20} + \alpha_{43}\beta_{30} + \beta_{40} \\
c_{41} &= \alpha_{43}\alpha_{32}\beta_{21} + \alpha_{42}\beta_{21} + \alpha_{43}\beta_{31} + \beta_{41} \\
c_{42} &= \alpha_{43}\beta_{32} + \beta_{42} \\
c_{43} &= \beta_{43}.
\end{aligned}
\tag{2.22}
$$

This relationship makes it clear that if all the α_{ij} and β_{ij} are non-negative, then the coefficients c_{ij} will also be non-negative.

Once again, we solve the order conditions up to fourth order and find that the coefficients must satisfy a two-parameter family or one of three special cases of a one-parameter family [83].

- **General Case.** If two parameters α_2 and α_3 are such that: $\alpha_2 \neq \alpha_3$, $\alpha_2 \neq 1$, $\alpha_2 \neq 0$, $\alpha_2 \neq \frac{1}{2}$, $\alpha_3 \neq 1$, $\alpha_3 \neq 0$, $\alpha_3 \neq \frac{1}{2}$, and $6\alpha_2\alpha_3 - 4(\alpha_2 + \alpha_3) + 3 \neq 0$. Then the coefficients c_{ij} are :

$c_{10} = \alpha_2$, $c_{20} = \alpha_3 - c_{21}$, $c_{21} = \frac{\alpha_3(\alpha_3 - \alpha_2)}{2\alpha_2(1 - 2\alpha_2)}$,

$c_{30} = 1 - c_{31} - c_{32}$,

$c_{31} = \frac{(1 - \alpha_2)[\alpha_2 + \alpha_3 - 1 - (2\alpha_3 - 1)^2]}{2\alpha_2(\alpha_3 - \alpha_2)[6\alpha_2\alpha_3 - 4(\alpha_2 + \alpha_3) + 3]}$,

$c_{32} = \frac{(1 - 2\alpha_2)(1 - \alpha_2)(1 - \alpha_3)}{\alpha_3(\alpha_3 - \alpha_2)[6\alpha_2\alpha_3 - 4(\alpha_2 + \alpha_3) + 3]}$,

$c_{40} = \frac{1}{2} + \frac{1 - 2(\alpha_2 + \alpha_3)}{12\alpha_2\alpha_3}$, $\quad c_{41} = \frac{2\alpha_3 - 1}{12\alpha_2(\alpha_3 - \alpha_2)(1 - \alpha_2)}$, $\quad c_{42} = $

$\frac{(1 - 2\alpha_2)}{12\alpha_3(\alpha_3 - \alpha_2)(1 - \alpha_3)}$, $c_{43} = \frac{1}{2} + \frac{2(\alpha_2 + \alpha_3) - 3}{12(1 - \alpha_2)(1 - \alpha_3)}$.

- **Special Case I.** If $\alpha_2 = \alpha_3$ the method can be fourth order only if $\alpha_2 = \alpha_3 = \frac{1}{2}$. In this case $c_{10} = \frac{1}{2}$, $c_{20} = \frac{1}{2} - \frac{1}{6w_3}$, $c_{21} = \frac{1}{6w_3}$, $c_{30} = 0$, $c_{31} = 1 - 3w_3$, $c_{32} = 3w_3$, $c_{40} = \frac{1}{6}$, $c_{41} = \frac{2}{3} - w_3$, $c_{42} = w_3$, $c_{43} = \frac{1}{6}$.

- **Special Case II.** If $\alpha_2 = 1$, the method can be fourth order only if $\alpha_3 = \frac{1}{2}$. Then $c_{10} = 1$, $c_{20} = \frac{3}{8}$, $c_{21} = \frac{1}{8}$, $c_{30} = 1 - c_{31} - c_{32}$, $c_{31} = -\frac{1}{12w_4}$, $c_{32} = \frac{1}{3w_4}$, $c_{40} = \frac{1}{6}$, $c_{41} = \frac{1}{6} - w_4$, $c_{42} = \frac{2}{3}$, $c_{43} = w_4$.

- **Special Case III.** If $\alpha_3 = 0$ the method can be fourth order only if $\alpha_2 = \frac{1}{2}$. Then $c_{10} = \frac{1}{2}$, $c_{20} = -\frac{1}{12w_3}$, $c_{21} = \frac{1}{12w_3}$, $c_{30} = 1 - c_{31} - c_{32}$, $c_{31} = \frac{3}{2}$, $c_{32} = 6w_3$, $c_{40} = \frac{1}{6} - w_3$, $c_{41} = \frac{2}{3}$, $c_{42} = w_3$, $c_{43} = \frac{1}{6}$.

Theorem 2.4. *If we require $\beta_{ij} \geq 0$, any four-stage fourth order SSP Runge–Kutta method (2.10) will not have $\mathcal{C} > 0$.*

Proof. We consider the coefficients above and show, for each case, that the requirement that all the Shu-Osher coefficients must be non-negative implies that the coefficients c_{ij} must be non-negative as well. In the following we show that if $\mathcal{C} > 0$, the coefficients cannot be non-negative.

- **General Case:** There are five possibilities to consider:

  (1) $\alpha_2 < 0$ implies $c_{10} < 0$.

  (2) $\alpha_3 > \alpha_2 > 0$ and $0 < \alpha_2 < \frac{1}{2}$:

  $c_{41} \geq 0$ requires $\alpha_3 > \frac{1}{2}$. $c_{20} \geq 0$ requires $\alpha_3 \leq 3\alpha_2 - 4\alpha_2^2 \leq \frac{9}{16}$. $c_{32} \geq 0$ and $c_{31} \geq 0$ require that $\alpha_2 \geq 2 - 5\alpha_3 + 4\alpha_3^2$. Since this is a decreasing function of α_3 when $\alpha_3 \leq \frac{9}{16}$, we obtain $\alpha_2 \geq 2 - 5(3\alpha_2 - 4\alpha_2^2) + 4(3\alpha_2 - 4\alpha_2^2)^2$. Rearranging, we find that $0 \geq 2((2\alpha_2 - 1)^2 + 4\alpha_2^2)(2\alpha_2 - 1)^2$, which is impossible.

  (3) $\alpha_3 < \alpha_2$ and $\alpha_2 > \frac{1}{2}$:

  $c_{42} \geq 0$ requires $0 < \alpha_3 < 1$.

  We can only have $c_{32} \geq 0$ in one of the two ways:

    (a) If $1 - \alpha_2 > 0$, and $6\alpha_2\alpha_3 - 4(\alpha_2 + \alpha_3) + 3 > 0$,
    we note that $c_{41} \geq 0$ requires $\alpha_3 < \frac{1}{2}$.

Simple calculation yields

$$c_{30} = 1 - c_{31} - c_{32}$$

$$= \frac{(2 - 6\alpha_2 + 4\alpha_2^2) + (-5 + 15\alpha_2 - 12\alpha_2^2)\alpha_3}{2\alpha_2\alpha_3(6\alpha_2\alpha_3 - 4(\alpha_2 + \alpha_3) + 3)}$$

$$+ \frac{(4 - 12\alpha_2 + 12\alpha_2^2)\alpha_3^2}{2\alpha_2\alpha_3(6\alpha_2\alpha_3 - 4(\alpha_2 + \alpha_3) + 3)},$$

hence $c_{30} \geq 0$ requires

$$A + B\alpha_3 + C\alpha_3^2 \equiv (2 - 6\alpha_2 + 4\alpha_2^2)$$
$$+ (-5 + 15\alpha_2 - 12\alpha_2^2)\alpha_3$$
$$+ (4 - 12\alpha_2 + 12\alpha_2^2)\alpha_3^2 \geq 0.$$

It is easy to show that, when $\frac{1}{2} < \alpha_2 < 1$, we have $A < 0$, $B < 0$ and $C > 0$. Thus, for $0 < \alpha_3 < \frac{1}{2}$, we have

$$A + B\alpha_3 + C\alpha_3^2 < \max\left(A, A + \frac{1}{2}B + \frac{1}{4}C\right)$$

$$= \max\left(A, \frac{1}{2}(1 - 2\alpha_2)(1 - \alpha_2)\right)$$

$$< 0$$

which is a contradiction.

(b) $\alpha_2 > 1$, and $6\alpha_2\alpha_3 - 4(\alpha_2 + \alpha_3) + 3 < 0$.

$c_{31} \geq 0$ requires $\alpha_2 + \alpha_3 - 1 - (2\alpha_3 - 1)^2 \leq 0$, which implies $(1 - 4\alpha_3)(1 - \alpha_3) = 4\alpha_3^2 - 5\alpha_3 + 1 \geq \alpha_2 - 1 > 0$. Clearly, this is true only if $\alpha_3 < \frac{1}{4}$.

Now, $c_{40} \geq 0$ requires that $0 \leq 6\alpha_2\alpha_3 - 2(\alpha_2 + \alpha_3) + 1 = 2\alpha_3(3\alpha_2 - 1) + (1 - 2\alpha_2) \leq \frac{1}{2}(3\alpha_2 - 1) + (1 - 2\alpha_2) = \frac{1}{2}(1 - \alpha_2)$, an apparent contradiction.

(4) $0 < \alpha_2 < \frac{1}{2}$ and $\alpha_3 < \alpha_2$: in this case we can see immediately that $c_{42} < 0$.

(5) If $\frac{1}{2} < \alpha_2 < \alpha_3$, $c_{21} < 0$.

- If $6\alpha_2\alpha_3 - 4(\alpha_2 + \alpha_3) + 3 = 0$, or if $\alpha_2 = 0$ or if $\alpha_3 = 1$, then this method is not fourth order [83].

- **Special Case I.** Clearly we need to have $c_{42} = w_3 \geq 0$. To have $c_{31} = 1 - 3w_3 \geq 0$ and $c_{20} = \frac{1}{2} - \frac{1}{6w_3} \geq 0$, we require $w_3 = \frac{1}{3}$. This leads to the classical fourth order Runge–Kutta method. Clearly, then, $\alpha_{21} = \frac{c_{20} - \beta_{20}}{\beta_{10}} = -2\beta_{20}$. This is only acceptable if $\alpha_{21} = \beta_{20} = 0$. But $\beta_{21} = \frac{1}{2}$, so in the case where all β_{ij}s are non-negative, the SSP coefficient (2.11) is equal to zero.

- **Special Case II.** In this case we want $c_{31} = -\frac{1}{12w_4} \geq 0$ which means $w_4 < 0$. But then $c_{43} = w_4 < 0$. So this case does not allow all non-negative β_{ij}s.
- **Special Case III.** Clearly, $c_{20} = -\frac{1}{12w_3} = -c_{21}$, one of these must be negative. Thus, this case does not allow all non-negative β_{ij}s, either.

$\square$

As we noted above, the SSP property can still be guaranteed with negative β_{ij}s as long as we adapt the spatial discretization (by downwinding) in those cases. Such methods will be addressed in Chapter 10.

The good news is that one can still obtain fourth order explicit SSP methods without downwinding if we add stages. Spiteri and Ruuth [98, 97] developed fourth order SSP methods with five through eight stages. The most popular fourth order method is the five-stage method which was found in [62] (in another context) and again independently in [97].

SSPRK(5,4):

$$u^{(1)} = u^n + 0.391752226571890\Delta t F(u^n)$$
$$u^{(2)} = 0.444370493651235u^n + 0.555629506348765u^{(1)}$$
$$+0.368410593050371\Delta t F(u^{(1)})$$
$$u^{(3)} = 0.620101851488403u^n + 0.379898148511597u^{(2)}$$
$$+0.251891774271694\Delta t F(u^{(2)})$$
$$u^{(4)} = 0.178079954393132u^n + 0.821920045606868u^{(3)}$$
$$0.544974750228521\Delta t F(u^{(3)})$$
$$u^{n+1} = 0.517231671970585u^{(2)}$$
$$+0.096059710526147u^{(3)} + 0.063692468666290\Delta t F(u^{(3)})$$
$$+0.386708617503269u^{(4)} + 0.226007483236906\Delta t F(u^{(4)}) .$$

This method has SSP coefficient $\mathcal{C} = 1.508$, and effective SSP coefficient $\mathcal{C}_{\text{eff}} = 0.302$, which means that this method is higher order and only slightly less efficient than the popular SSPRK(3,3). This method was numerically found to be optimal.

If we are willing to add more stages, we can obtain a method (developed in [55]) with twice as large a value of $\mathcal{C}_{\text{eff}}$. This method also has rational coefficients and a low-storage implementation.

SSPRK(10,4):

$$u^{(1)} = u^n + \frac{1}{6}\Delta t F(u^n)$$

$$u^{(2)} = u^{(1)} + \frac{1}{6}\Delta t F(u^{(1)})$$

$$u^{(3)} = u^{(2)} + \frac{1}{6}\Delta t F(u^{(2)})$$

$$u^{(4)} = u^{(3)} + \frac{1}{6}\Delta t F(u^{(3)})$$

$$u^{(5)} = \frac{3}{5}u^n + \frac{2}{5}u^{(4)} + \frac{1}{15}\Delta t F(u^{(4)})$$

$$u^{(6)} = u^{(5)} + \frac{1}{6}\Delta t F(u^{(5)})$$

$$u^{(7)} = u^{(6)} + \frac{1}{6}\Delta t F(u^{(6)})$$

$$u^{(8)} = u^{(7)} + \frac{1}{6}\Delta t F(u^{(7)})$$

$$u^{(9)} = u^{(8)} + \frac{1}{6}\Delta t F(u^{(8)})$$

$$u^{n+1} = \frac{1}{25}u^n + \frac{9}{25}u^{(4)} + \frac{3}{5}u^{(9)} + \frac{3}{50}\Delta t F(u^{(4)}) + \frac{1}{10}\Delta t F(u^{(9)}).$$

This method has SSP coefficient $\mathcal{C} = 6$, and effective SSP coefficient $\mathcal{C}_{\text{eff}} = 3/5$. It is optimal among all fourth order, ten-stage methods. As we see from the effective SSP coefficient, the cost of additional stages is more than offset by the larger SSP coefficient. In Chapter 6 we give a low-storage implementation of this method. The issue of low storage methods will be discussed in that chapter, where we will present methods of orders two and three which are optimal in terms of their SSP coefficient as well as low storage properties.

Adding stages allows us to find methods with a larger effective SSP coefficient, and fourth order SSP methods, which are impossible for a four-stage method but possible with more stages. Unfortunately, adding stages can only take us so far. It will not allow us to obtain higher order SSP methods. We will see in Chapter 5 that explicit SSP Runge–Kutta methods without downwinding cannot have a positive SSP coefficient and be greater than fourth order accurate.

Chapter 3

The SSP Coefficient for Runge–Kutta Methods

In Chapter 2 we referred to the SSP coefficient $\mathcal{C}(\alpha, \beta)$ as if it were a well-defined property of a Runge–Kutta method. But as we showed in Example 2.2, by writing the same method in different but equivalent Shu-Osher forms, we can obtain different values of $\mathcal{C}(\alpha, \beta)$. In the following example we show that it is possible to obtain an infinite number of values of $\mathcal{C}(\alpha, \beta)$ from the same method.

Example 3.1. Recall the second order Runge–Kutta method, based on the trapezoidal rule, discussed in Example 2.2:

$$u^{(1)} = u^n + \Delta t F(u^n) \tag{3.1a}$$

$$u^{n+1} = u^n + \frac{1}{2}\Delta t F(u^n) + \frac{1}{2}\Delta t F(u^{(1)}). \tag{3.1b}$$

In this form, we obtain $\mathcal{C}(\alpha, \beta) = 0$ because $\alpha_{32} = 0$ while $\beta_{32} = 1/2$. However, using equation (3.1a), we can rewrite (3.1b) to obtain a positive value of $\mathcal{C}(\alpha, \beta)$. For instance, (3.1b) can be written in the form (2.14)

$$u^{n+1} = \frac{3}{4}u^n + \frac{1}{4}\Delta t F(u^n) + \frac{1}{4}u^{(1)} + \frac{1}{2}\Delta t F(u^{(1)}), \tag{3.2}$$

which yields $\mathcal{C}(\alpha, \beta) = 1/2$. This is still not optimal; (3.1b) is also equivalent to (2.15)

$$u^{n+1} = \frac{1}{2}u^n + \frac{1}{2}u^{(1)} + \frac{1}{2}\Delta t F(u^{(1)}), \tag{3.3}$$

which yields $\mathcal{C}(\alpha, \beta) = 1$. It turns out that this is the largest possible value of $\mathcal{C}(\alpha, \beta)$ for this method. In fact, it is possible to obtain any value of $\mathcal{C}(\alpha, \beta) \leq 1$ by multiplying (3.1a) by any constant $0 \leq \hat{\alpha} \leq \frac{1}{2}$ and adding zero in the form

$$\hat{\alpha}u^{(1)} - \hat{\alpha}\left(u^n + \Delta t F(u^n)\right)$$

to (3.1b) to obtain

$$u^{n+1} = (1 - \hat{\alpha}) \left(u^n + \frac{1 - 2\hat{\alpha}}{2 - 2\hat{\alpha}} \Delta t F(u^n) \right) + \hat{\alpha} \left(u^{(1)} + \frac{1}{2\hat{\alpha}} \Delta t F(u^{(1)}) \right)$$

$$(3.4)$$

which has $\mathcal{C}(\alpha, \beta) = 2\hat{\alpha}$. $\qquad \Box$

In addition to the fact that a given Runge–Kutta method can be written in infinitely many Shu-Osher forms, some methods can also be written in a form that uses fewer stages than the original representation. Such methods are said to be *reducible*.

Example 3.2. Consider the method

$$u^{(1)} = u^n - \Delta t F(u^n) \qquad (3.5a)$$

$$u^{n+1} = 2u^n - u^{(1)}. \qquad (3.5b)$$

For this method, we obtain $\mathcal{C}(\alpha, \beta) = 0$, since some of the coefficients are negative. However, this method is reducible: (3.5b) can be rewritten as

$$u^{n+1} = u^n + \Delta t F(u^n),$$

which is just the forward Euler method, with $\mathcal{C}(\alpha, \beta) = 1$. Note that the stage $u^{(1)}$ need not be computed at all. $\qquad \Box$

Given these ambiguities – that a method may be equivalent to one with fewer stages and that different values of $\mathcal{C}(\alpha, \beta)$ may be obtained for a given method – we will in this chapter refer to $\mathcal{C}(\alpha, \beta)$ as an *apparent SSP coefficient*. What we really want to know is not the apparent SSP coefficient, but the best apparent SSP coefficient we can obtain for a given method. That is, we are interested in the maximal value

$$\mathcal{C} = \max_{\alpha, \beta} \mathcal{C}(\alpha, \beta) \qquad (3.6)$$

where the maximum is taken over all representations α, β corresponding to a given method. We refer to $\mathcal{C}$ as the *strong stability preserving coefficient* of the method. In this chapter, we show how to determine $\mathcal{C}$.

The remainder of the chapter proceeds as follows. First, we introduce the full class of implicit Runge–Kutta methods in Section 3.1. Second, we need a way to unambiguously refer to a particular Runge–Kutta method, independent of its representation. This will necessitate a discussion of the Butcher form and of reducibility, in Section 3.2. In Section 3.3, we present a unique form that reveals the value of $\mathcal{C}$ for any method. In Section 3.4, we formulate the problem of finding optimal SSP Runge–Kutta methods. In Section 3.5, we demonstrate by a constructive proof that the SSP time step restriction is necessary for the preservation of strong stability bounds.

3.1   The modified Shu-Osher form

The methods we have considered thus far are known as *explicit* methods, since each stage $u^{(i)}$ is a function of u^n and of the previous stages $u^{(1)}, u^{(2)}, \ldots, u^{(i-1)}$. In general, we can consider methods in which each stage $u^{(i)}$ depends on all the other stages $u^{(1)}, u^{(2)}, \ldots, u^{(m)}$. Such methods are referred to as *implicit* methods. The following generalization of the Shu-Osher form can be used to represent both explicit and implicit methods:

$$u^{(i)} = v_i u^n + \sum_{j=1}^{m} \left(\alpha_{ij} u^{(j)} + \Delta t \beta_{ij} F(u^{(j)}) \right) \quad (1 \le i \le m+1) \quad (3.7)$$

$$u^{n+1} = u^{(m+1)}.$$

This form was first defined in [21, 37] and is referred to as the modified Shu-Osher form. It differs from the original Shu-Osher form (2.10) in three ways. First, implicit terms are included, and second, terms that are proportional to u^n are written explicitly, since implicit methods do not generally have the first stage equal to u^n. Finally, the stages are indexed starting from one rather than zero in order to agree with the stage numbering in Butcher form.

Let us rearrange (3.7) as follows:

$$u^{(i)} = v_i u^n + \sum_{j=1}^{m} \alpha_{ij} \left(u^{(j)} + \Delta t \frac{\beta_{ij}}{\alpha_{i,j}} F(u^{(j)}) \right) \quad (1 \le i \le m+1). \quad (3.8)$$

Since consistency requires that

$$v_i + \sum_{j=1}^{m} \alpha_{ij} = 1 \qquad\qquad 1 \le i \le m+1, \qquad (3.9)$$

then if $\alpha_{ij}, \beta_{ij}, v_i \ge 0$, each stage $u^{(i)}$ is a convex combination of forward Euler steps. This suggests a way to generalize Theorem 2.1 to implicit methods.

However, we must exclude certain defective methods. We will say a Runge–Kutta method is zero-well-defined if the stage equations have a unique solution when the method is applied to the scalar initial value problem (IVP)

$$u'(t) = 0, \qquad u(t_0) = u_0. \qquad (3.10)$$

Example 3.3. Consider a two-stage method given in the modified Shu-Osher form (3.7)

$$u^{(1)} = u^{(2)} + f(u^{(1)})$$
$$u^{(2)} = u^{(1)} + f(u^{(1)})$$
$$u^{n+1} = u^{(2)}.$$

Taking $f(u) = 0$, the stage equations both reduce to $u^{(1)} = u^{(2)}$, so $u^{(1)}$ (and also u^{n+1}) is arbitrary. Hence this method is not zero-well-defined. $\square$

Clearly, methods that are not zero-well-defined are of no practical interest.

Theorem 3.1. *Suppose that F satisfies the forward Euler condition (2.6), and let u^n denote the solution at step n given by applying a zero-well-defined Runge–Kutta method (3.7) to the initial value problem (2.2). Then the solution satisfies the strong stability bound*

$$\|u^{n+1}\| \leq \|u^n\|$$

if the time step satisfies

$$0 \leq \Delta t \leq \mathcal{C}(\boldsymbol{\alpha}, \boldsymbol{\beta}) \Delta t_{\mathrm{FE}} \tag{3.11}$$

where now

$$\mathcal{C}(\boldsymbol{\alpha}, \boldsymbol{\beta}) = \begin{cases} \min_{i,j} \frac{\alpha_{ij}}{\beta_{ij}} & \text{if all } \alpha_{ij}, \beta_{ij}, v_i \text{ are non-negative} \\ 0 & \text{otherwise.} \end{cases}$$

As before, we consider the ratio $\frac{\alpha_{ij}}{\beta_{ij}}$ to be infinite if $\beta_{ij} = 0$.

Proof. If $\mathcal{C}(\boldsymbol{\alpha}, \boldsymbol{\beta}) = 0$, we must have $\Delta t = 0$, so that $u^{n+1} = u^n$, and the statement is trivial. Suppose $\mathcal{C}(\boldsymbol{\alpha}, \boldsymbol{\beta}) > 0$. Taking norms on both sides of (3.8), and using (3.9), convexity of $\|\cdot\|$, and the forward Euler condition, we obtain the bound

$$
\begin{aligned}
\left\| u^{(i)} \right\| &= \left\| \left(1 - \sum_{j=1}^{m} \alpha_{ij} \right) u^n + \sum_{j=1}^{m} \alpha_{ij} u^{(j)} + \Delta t \frac{\beta_{ij}}{\alpha_{i,j}} F(u^{(j)}) \right\| \\
&\leq \left(1 - \sum_{j=1}^{m} \alpha_{ij} \right) \|u^n\| + \sum_{j=1}^{m} \alpha_{ij} \left\| u^{(j)} + \Delta t \frac{\beta_{ij}}{\alpha_{i,j}} F(u^{(j)}) \right\| \\
&\leq \left(1 - \sum_{j=1}^{m} \alpha_{ij} \right) \|u^n\| + \sum_{j=1}^{m} \alpha_{ij} \left\| u^{(j)} \right\|.
\end{aligned}
\tag{3.12}
$$

Now let q be the index of the Runge–Kutta stage with largest norm, i.e. choose $q \in \{1, 2, \ldots, m+1\}$ such that $\|u^{(i)}\| \le \|u^{(q)}\|$ for all $1 \le i \le m+1$. Then taking $i = q$ in (3.12) yields

$$\left\| u^{(q)} \right\| \le \left(1 - \sum_{j=1}^{m} \alpha_{qj} \right) \|u^n\| + \sum_{j=1}^{m} \alpha_{qj} \left\| u^{(j)} \right\|$$

$$\left\| u^{(q)} \right\| \le \left(1 - \sum_{j=1}^{m} \alpha_{qj} \right) \|u^n\| + \sum_{j=1}^{m} \alpha_{qj} \left\| u^{(q)} \right\|$$

$$\left(1 - \sum_{j=1}^{m} \alpha_{qj} \right) \left\| u^{(q)} \right\| \le \left(1 - \sum_{j=1}^{m} \alpha_{qj} \right) \|u^n\|$$

$$\left\| u^{(J)} \right\| \le \|u^n\|.$$

Here we have assumed $1 - \sum_j \alpha_{qj} \ne 0$. Suppose instead that $1 - \sum_j \alpha_{qj} = 0$. Then from (3.12) we obtain

$$\left\| u^{(q)} \right\| \le \sum_{j=1}^{m} \alpha_{qj} \left\| u^{(j)} \right\|.$$

Since q was chosen so that $\|u^{(i)}\| \le \|u^{(q)}\|$ for all $1 \le i \le m+1$, this implies that $\|u^{(j)}\| = \|u^{(q)}\|$ for every j such that $\alpha_{qj} \ne 0$. Let

$$J = \{j : \alpha_{qj} \ne 0\}.$$

If there exists any $j^* \in J$ such that $1 - \sum_j \alpha_{qj} \ne 0$, we can take $q = j^*$ and apply the argument above. If not, then it follows that the stages with indices in J depend only on each other, and not on u^n. Then the method is not zero-well-defined, since these stages can take any of infinitely many possible values when the method is applied to (3.10). $\qquad\square$

3.1.1 Vector notation

In the following sections, it will be helpful to represent a Runge–Kutta method using a more compact notation. To this end, let us formalize the definition of the matrices $\boldsymbol{\alpha}, \boldsymbol{\beta}$:

$$(\boldsymbol{\alpha})_{ij} = \begin{cases} \alpha_{ij} & 1 \le i \le m+1, 1 \le j \le m \\ 0 & j = m+1 \end{cases}$$

$$(\boldsymbol{\beta})_{ij} = \begin{cases} \beta_{ij} & 1 \le i \le m+1, 1 \le j \le m, \\ 0 & j = m+1. \end{cases}$$

To emphasize that the solution u^n and the intermediate stages $u^{(i)}$ are themselves usually vectors, we will use the notation $\mathbf{u}^n$ and $\mathbf{u}^{(i)}$ in much of the following discussion. For the same reason, we also define the vectors $\boldsymbol{y}, \boldsymbol{f}$ by

$$\boldsymbol{y}_i = \mathbf{u}^{(i)}$$
$$\boldsymbol{f}_i = F(\mathbf{u}^{(i)}).$$

As we have the Runge–Kutta method operating on vectors, we need to introduce the Kronecker-product matrices

$$\bar{\boldsymbol{v}} = \mathbf{I} \otimes \mathbf{v}$$
$$\bar{\boldsymbol{\alpha}} = \mathbf{I} \otimes \boldsymbol{\alpha}$$
$$\bar{\boldsymbol{\beta}} = \mathbf{I} \otimes \boldsymbol{\beta},$$

so that we can compactly write a Runge–Kutta method:

$$\boldsymbol{y} = \bar{\boldsymbol{v}} \mathbf{u}^n + \bar{\boldsymbol{\alpha}} \boldsymbol{y} + \Delta t \bar{\boldsymbol{\beta}} \boldsymbol{f} \tag{3.13}$$
$$\mathbf{u}^{n+1} = \boldsymbol{y}_{m+1}.$$

Example 3.4. Let us consider the trapezoidal Runge–Kutta method in Example 2.2. As we saw, it can be represented as

$$\mathbf{u}^{(1)} = \mathbf{u}^n \tag{3.14a}$$
$$\mathbf{u}^{(2)} = \mathbf{u}^{(1)} + \Delta t F(\mathbf{u}^{(1)}) \tag{3.14b}$$
$$\mathbf{u}^{n+1} = \frac{3}{4}\mathbf{u}^{(1)} + \frac{1}{4}\Delta t F(\mathbf{u}^{(1)}) + \frac{1}{4}\mathbf{u}^{(2)} + \frac{1}{2}\Delta t F(\mathbf{u}^{(2)}). \tag{3.14c}$$

This is a modified Shu-Osher form with

$$\boldsymbol{\beta} = \begin{pmatrix} 0 & 0 \\ 1 & 0 \\ \frac{1}{4} & \frac{1}{2} \end{pmatrix}, \qquad \boldsymbol{\alpha} = \begin{pmatrix} 0 & 0 \\ 1 & 0 \\ \frac{3}{4} & \frac{1}{4} \end{pmatrix}, \qquad \mathbf{v} = \begin{pmatrix} 1 \\ 0 \\ 0 \end{pmatrix}. \tag{3.15}$$

$\square$

3.2 Unique representations

In order to discuss the maximum SSP coefficient over all representations of a Runge–Kutta method, we need a way of uniquely referring to a particular method. The Shu-Osher form (2.10) and modified Shu-Osher form (3.7) are very convenient for SSP analysis, but as we have noted, they are not unique.

3.2.1 The Butcher form

One way to eliminate this ambiguity is to take $\alpha = 0$ in the modified Shu-Osher form (3.7). By (3.9), we have that $\mathbf{v} = \mathbf{e}$ in this case. What about the coefficients β_{ij}? For the moment, we simply denote the resulting array of coefficients β_{ij} by β_0. Thus the method is

$$y = \mathbf{e}u^n + \Delta t \bar{\beta}_0 f \tag{3.16}$$
$$u^{n+1} = y_{m+1}.$$

To find the relation between the modified Shu-Osher coefficients and the Butcher coefficients, we simply solve (3.13) for y (assuming for the moment that $I - \alpha$ is invertible):

$$y = \bar{v}u^n + \bar{\alpha}y + \Delta t \bar{\beta} f \tag{3.17}$$
$$(I - \bar{\alpha})y = \bar{v}u^n + \Delta t \bar{\beta} f$$
$$y = (I - \bar{\alpha})^{-1} \bar{v}u^n + \Delta t (I - \bar{\alpha})^{-1} \bar{\beta} f$$
$$y = \mathbf{e}u^n + \Delta t (I - \bar{\alpha})^{-1} \bar{\beta} f. \tag{3.18}$$

Here we have used the relation (3.9), which is just $(I - \alpha)^{-1} \mathbf{v} = \mathbf{e}$. Comparing (3.18) with (3.16), we see that

$$\beta_0 = (I - \alpha)^{-1} \beta. \tag{3.19}$$

In fact, the form (3.16) is the Butcher form (2.9):

$$\mathbf{u}^{(i)} = \mathbf{u}^n + \Delta t \sum_{j=1}^{m} a_{ij} F(\mathbf{u}^{(j)}) \quad (1 \le i \le m) \tag{3.20}$$

$$\mathbf{u}^{n+1} = \mathbf{u}^n + \Delta t \sum_{j=1}^{m} b_j F(\mathbf{u}^{(j)}),$$

with

$$\beta_0 = \begin{pmatrix} \mathbf{A} & 0 \\ b^{\mathrm{T}} & 0 \end{pmatrix}. \tag{3.21}$$

The Butcher form of a given method *is* unique, unless the method happens to be reducible (a discussion of reducibility appears in Section 3.2.2).

What if $I - \alpha$ is singular? It turns out that this holds only for a trivial class of methods.

Lemma 3.1. *Let α, β be the coefficients of a Runge–Kutta method in the modified Shu-Osher form (3.7). The following three assertions are equivalent:*

(1) $\mathbf{I} - \boldsymbol{\alpha}$ *is invertible.*

(2) The method can be written in Butcher form.

(3) The method is zero-well-defined.

Proof. We will show that (1) $\implies$ (2), (2) $\implies$ (3), and (3) $\implies$ (1). From (3.19) we see immediately that (1) $\implies$ (2).

To see that (2) implies (3), assume the method can be written in Butcher form, and apply it to the trivial IVP (3.10) in this form. Then (3.16) simplifies to

$$\mathbf{y} = u^n \mathbf{e}.$$

Hence the method is zero-well-defined.

Finally, to see that (3) implies (1), apply the method (in modified Shu-Osher form) to (3.10). Then (3.13) simplifies to

$$(\mathbf{I} - \boldsymbol{\alpha})\mathbf{y} = u^n \mathbf{v}.$$

If $\mathbf{I} - \boldsymbol{\alpha}$ is singular, then this equation has infinitely many solutions (since $\mathbf{v} = (\mathbf{I} - \boldsymbol{\alpha})\mathbf{e}$). $\qquad\square$

3.2.2 *Reducibility of Runge–Kutta methods*

Even using the Butcher form, it is possible to represent a single method in multiple equivalent ways. Of course, a different representation can be obtained by permuting the stages, but this is not important for our analysis. A more significant source of ambiguity is that a Butcher form with m stages may be equivalent to a method with fewer than m stages. Such a method is said to be *reducible*. Runge–Kutta methods may be reducible in either of two distinct ways; we will be concerned only with what is known as *DJ-reducibility*.

Definition 3.1 (DJ-reducibility). *If $\alpha_{ij} \neq 0$ or $\beta_{ij} \neq 0$, we say that stage j influences stage i directly. If there are indices $i_1, i_2, ..., i_m$ such that stage i_k influences stage i_{k+1} for $1 \leq k \leq m - 1$, we say that stage i_1 influences stage i_m. Clearly, if some stage j does not influence stage $m+1$, then u^{n+1} may be computed without computing $u^{(j)}$, so the method can be written equivalently by removing the jth row and jth column of $\boldsymbol{\alpha}$ and the jth component of $\mathbf{v}$. If such a superfluous stage exists, we say the method is DJ-reducible; otherwise it is DJ-irreducible.*

Example 3.5. Consider a method with

$$\mathbf{v} = \begin{pmatrix} 1 \\ 0 \\ 0 \end{pmatrix}, \qquad \boldsymbol{\alpha} = \boldsymbol{\beta} = \begin{pmatrix} 0 & 0 & 0 \\ \frac{1}{2} & \frac{1}{2} & 0 \\ 1 & 0 & 0 \end{pmatrix}. \tag{3.22}$$

Since $\alpha_{32} = \beta_{32} = 0$, the second stage is irrelevant and the method is equivalent to the one given by removing the second row and column of $\boldsymbol{\alpha}$ and the second row of $\mathbf{v}$:

$$\mathbf{v} = \begin{pmatrix} 1 \\ 0 \end{pmatrix}, \qquad \boldsymbol{\alpha} = \begin{pmatrix} 0 & 0 \\ 1 & 0 \end{pmatrix}. \tag{3.23}$$

$\square$

Clearly, restricting our discussion to irreducible methods is no real restriction; we are merely demanding that the method be written in a sensible way.

3.3   The canonical Shu-Osher form

The Shu-Osher and modified Shu-Osher forms are useful for SSP analysis, but they are not unique. The Butcher form is unique for irreducible methods, but does not reveal the SSP coefficient. In this section we present a form that is both unique for irreducible Runge–Kutta methods and makes apparent the true SSP coefficient (3.6). This form turns out to be closely related to the radius of absolute monotonicity of a Runge–Kutta method [62] (see also [19, 37, 96]).

To find the SSP coefficient of a Runge–Kutta method, it is useful to consider a particular modified Shu-Osher form, in which the ratio $r = \alpha_{ij}/\beta_{ij}$ is the same for every i, j such that $\beta_{ij} \neq 0$. In other words, we write the method as a convex combination of forward Euler steps, where every step has the same length.

We will denote the coefficient matrices of this special form by $\boldsymbol{\alpha}_r, \boldsymbol{\beta}_r$, and require that $\boldsymbol{\alpha}_r = r\boldsymbol{\beta}_r$. Substituting this relation into (3.19), we can solve for $\boldsymbol{\beta}_r$ in terms of $\boldsymbol{\beta}_0$ and r. Assuming the method is zero-well-defined, we find

$$(\mathbf{I} - r\boldsymbol{\beta}_r)^{-1} \boldsymbol{\beta}_r = \boldsymbol{\beta}_0$$
$$\boldsymbol{\beta}_r = \boldsymbol{\beta}_0 - r\boldsymbol{\beta}_r\boldsymbol{\beta}_0$$
$$\boldsymbol{\beta}_r (\mathbf{I} + r\boldsymbol{\beta}_0) = \boldsymbol{\beta}_0.$$

Hence, if $\mathbf{I} + r\boldsymbol{\beta}_0$ is invertible, the coefficients for this form are given by

$$\boldsymbol{\beta}_r = \boldsymbol{\beta}_0 \left(\mathbf{I} + r\boldsymbol{\beta}_0\right)^{-1} \tag{3.24a}$$

$$\boldsymbol{\alpha}_r = r\boldsymbol{\beta}_r = r\boldsymbol{\beta}_0 \left(\mathbf{I} + r\boldsymbol{\beta}_0\right)^{-1} \tag{3.24b}$$

$$\mathbf{v}_r = (\mathbf{I} - \boldsymbol{\alpha}_r)\mathbf{e} = \left(\mathbf{I} + r\boldsymbol{\beta}_0\right)^{-1}\mathbf{e}. \tag{3.24c}$$

We will refer to the form given by the coefficients (3.24a)-(3.24c) as a *canonical Shu-Osher form*:

$$y = \bar{v}_r \mathbf{u}^{n-1} + \bar{\alpha}_r \left(y + \frac{\Delta t}{r} f\right). \tag{3.25}$$

Note also that the Butcher form (3.16), with coefficient matrix $\boldsymbol{\beta}_0$, corresponds to the canonical Shu-Osher form with $r = 0$. Program 3.1 is a MATLAB function for transforming from the Butcher form to the canonical Shu-Osher form for a given value of r.

Program 3.1 (Butcher form to canonical Shu-Osher form).
Generates the canonical Shu-Osher form of an explicit Runge–Kutta method, given its Butcher form and radius of absolute monotonicity.

Inputs: Butcher matrix **A** *and vector* **b**, *and scalar* r, *and number of stages* m.
Output: coefficients of the canonical Shu-Osher form $\boldsymbol{\alpha}_r, \boldsymbol{\beta}_r$.

```
function [alpha,beta] = canonical_shu_osher(A,b,m,r)
beta_0 = [A;b'];
beta = beta_0 / (eye(m) + r*A)
alpha = r*beta
```

Since $\mathcal{C}(\boldsymbol{\alpha}_r, \boldsymbol{\beta}_r) = r$ as long as the coefficients given by (3.24a)-(3.24c) are non-negative, we now have a way to construct a modified Shu-Osher form with apparent SSP coefficient r as long as

$$\left(\mathbf{I} + r\boldsymbol{\beta}_0\right)^{-1} \text{ exists, } \boldsymbol{\alpha}_r \geq 0, \text{ and } \mathbf{v}_r \geq 0. \tag{3.26}$$

Hence the SSP coefficient of the method with Butcher form $\boldsymbol{\beta}_0$ is at least r if these conditions hold.

In fact, the following theorem tells us that the SSP coefficient is equal to the maximum value of r such that (3.26) holds. In other words, **a Runge–Kutta method with SSP coefficient $\mathcal{C}$ can always be written in the canonical Shu-Osher form above with $r = \mathcal{C}$.**

Theorem 3.2. *Consider a Runge–Kutta method with Butcher coefficient array β_0. The SSP coefficient of the method is*

$$\mathcal{C} = \max\{r \geq 0 \mid (\mathbf{I} + r\beta_0)^{-1} \text{ exists, } \alpha_r \geq 0, \text{ and } \mathbf{v}_r \geq 0\}. \qquad (3.27)$$

Here the inequalities are meant componentwise.

The statement of this theorem achieves our goal: we have an expression for the SSP coefficient in terms of an unambiguous representation of a Runge–Kutta method. The proof of the theorem relies on two lemmas; these lemmas will also indicate a straightforward method to compute $\mathcal{C}$.

Before stating and proving the lemmas, we remark that the quantity defined by (3.27) was originally introduced (in a different, but equivalent form) in [62]. In that and other related works, it is referred to as the radius of absolute monotonicity of the Runge–Kutta method, due to its relation to the radius of absolute monotonicity of certain functions (see Lemma 3.3 below).

In addition to its theoretical usefulness, the canonical form with $r = \mathcal{C}$ is often advantageous for implementation. For many optimal SSP methods, this form results in very sparse coefficient matrices $\alpha_{\mathcal{C}}, \beta_{\mathcal{C}}$ so that the method can be implemented using very little memory. This will be examined in detail in Chapter 6.

We now present two technical lemmas needed for the proof of Theorem 3.2.

Lemma 3.2. *Let α, β be Shu-Osher arrays of a zero-well-defined method, and let β_0 be the Butcher array of the method. Then $\mathbf{I} + r\beta_0$ is invertible for all r in the range $0 \leq r \leq \mathcal{C}(\alpha, \beta)$.*

Proof. Since the method is zero-well-defined, Lemma 3.1 implies that $\mathbf{I} - \alpha$ is invertible. Since

$$\begin{aligned} \mathbf{I} + r\beta_0 &= \mathbf{I} + r(\mathbf{I} - \alpha)^{-1}\beta \\ &= (\mathbf{I} - \alpha)^{-1}(\mathbf{I} - \alpha + r\beta), \end{aligned}$$

then it suffices to prove that $\mathbf{I} - (\alpha - r\beta)$ is invertible. For $r = 0$, this follows from Lemma 3.1. Assume $r > 0$; then $\mathcal{C}(\alpha, \beta) > 0$ so we have $\alpha \geq 0$, $\beta \geq 0$, $\alpha - r\beta \geq 0$, and $\alpha e \leq e$. Together these imply that

$$(\alpha - r\beta)e \leq e. \qquad (3.28)$$

Suppose that $\mathbf{I} - (\alpha - r\beta)$ is singular; then, setting $\gamma = \alpha - r\beta$, we know $\gamma \mathbf{x} = \mathbf{x}$ for some vector $\mathbf{x}$. Let x_i denote the entry of $\mathbf{x}$ with largest modulus. Then

$$\sum_{j\neq i}\gamma_{ij}x_j = (1-\gamma_{ii})x_i. \tag{3.29}$$

Taking the modulus on both sides, we have

$$(1-\gamma_{ii})|x_i| = \left|\sum_{j\neq i}\gamma_{ij}x_j\right| \leq \sum_{j\neq i}\gamma_{ij}|x_j|. \tag{3.30}$$

Dividing by $|x_i|$ gives

$$1-\gamma_{ii} \leq \sum_{j\neq i}\gamma_{ij}\frac{|x_j|}{|x_i|} \leq \sum_{j\neq i}\gamma_{ij} \leq 1-\gamma_{ii}. \tag{3.31}$$

Evidently all inequalities in this last equation are equalities. This implies that

$$|x_j| = |x_i| \text{ for all } j.$$

Combining this with (3.28) implies that $\mathbf{x} = \mathbf{e}$, so $\boldsymbol{\gamma}\mathbf{e} = \mathbf{e}$. Thus we have

$$\mathbf{e} = (\boldsymbol{\alpha} - r\boldsymbol{\beta})\mathbf{e} \leq \mathbf{e} - r\boldsymbol{\beta}\mathbf{e}.$$

Since $r > 0$ and $\boldsymbol{\beta} \geq 0$, this implies that $\boldsymbol{\beta} = 0$. But then Lemma 3.1 implies that $\mathbf{I} - \boldsymbol{\gamma}$ is invertible, which is a contradiction. $\qquad\square$

Lemma 3.3. *Let* $\boldsymbol{\alpha}, \boldsymbol{\beta}$ *be Shu-Osher arrays of a zero-well-defined method, and let* $\boldsymbol{\beta}_0$ *be the Butcher array of the method. For all* $0 \leq r \leq \mathcal{C}(\boldsymbol{\alpha}, \boldsymbol{\beta})$, *(3.26) holds.*

Proof. Take $0 \leq r < \mathcal{C}$. Since $r < \mathcal{C}(\boldsymbol{\alpha}_\mathcal{C}, \boldsymbol{\beta}_\mathcal{C})$, Lemma 3.2 implies that $\mathbf{I} + r\boldsymbol{\beta}_0$ is invertible. It remains to show that $\boldsymbol{\alpha}_r, \mathbf{v}_r \geq 0$.
    Define

$$S(z) = \boldsymbol{\beta}_0(\mathbf{I} - z\boldsymbol{\beta}_0)^{-1} \quad \text{and} \quad T(z) = (\mathbf{I} - z\boldsymbol{\beta}_0)^{-1}.$$

Since $S'(z) = (S(z))^2$ and $T'(z) = S(z)T(z)$, it follows by induction that

$$S^{(k)}(z) = k!(S(z))^{k+1} \quad \text{and} \quad T^{(k)}(z) = k!(S(z))^k T(z).$$

Hence we can write

$$\boldsymbol{\alpha}_r = r\boldsymbol{\beta}_0(\mathbf{I} + r\boldsymbol{\beta}_0)^{-1} = rS(-r) = r\sum_{k=0}^{\infty}\frac{(\mathcal{C}-r)^{k-1}}{k!}S^{(k)}(-\mathcal{C})$$

$$= r\sum_{k=0}^{\infty}(\mathcal{C}-r)^{k-1}S(-\mathcal{C})^k$$

$$= r\sum_{k=0}^{\infty}\frac{(\mathcal{C}-r)^{k-1}}{\mathcal{C}^k}\boldsymbol{\alpha}_\mathcal{C}^k \geq 0,$$

since $S(-\mathcal{C}) = \alpha_\mathcal{C}/\mathcal{C}$. Similarly,

$$\mathbf{v}_r = (\mathbf{I} + r\boldsymbol{\beta}_0)^{-1}\mathbf{e} = T(-r)\mathbf{e} = \sum_{k=0}^{\infty} \frac{(\mathcal{C} - r)^k}{k!} T^{(k)}(-\mathcal{C})\mathbf{e}$$

$$= \sum_{k=0}^{\infty} (\mathcal{C} - r)^k S^k(-\mathcal{C})T(-\mathcal{C})\mathbf{e}$$

$$= \sum_{k=0}^{\infty} \frac{(\mathcal{C} - r)^k}{\mathcal{C}^k} \alpha_\mathcal{C}^k \mathbf{v}_\mathcal{C} \geq 0.$$

$\square$

Theorem 3.2 follows easily from Lemmas 3.2 and 3.3. Suppose that (3.26) holds for some r; then $\mathcal{C} \geq \mathcal{C}(\boldsymbol{\alpha}_r, \boldsymbol{\beta}_r) = r$. On the other hand, suppose $r \in [0, \mathcal{C}]$. Then by Lemmas 3.2 and 3.3, (3.26) holds.

We can formulate the radius of absolute monotonicity in a slightly different way, as follows. Recalling that by (3.9) $\mathbf{v} = (\mathbf{I} - \boldsymbol{\alpha})\mathbf{e}$, we obtain the equivalence (for $r \geq 0$)

$$r \leq \mathcal{C} \iff (\mathbf{I} + r\boldsymbol{\beta}_0)^{-1} \text{ exists and } \|[\mathbf{v}_r \quad \boldsymbol{\alpha}_r]\|_\infty \leq 1, \tag{3.32}$$

where $[\mathbf{v}_r \quad \boldsymbol{\alpha}_r]$ is the matrix formed by adjoining $\mathbf{v}_r$ and $\boldsymbol{\alpha}_r$.

3.3.1 *Computing the SSP coefficient*

Lemmas 3.2 and 3.3 imply that we can construct a canonical Shu-Osher form with apparent SSP coefficient r for any $r \in [0, \mathcal{C}]$, whereas Theorem 3.2 implies that no such form exists for $r > \mathcal{C}$. Thus, given any method, we can compute its SSP coefficient by bisection. For any value of r, we simply construct the corresponding canonical form and check whether the resulting coefficients $\boldsymbol{\alpha}_\mathcal{C}, \mathbf{v}_\mathcal{C}$ are non-negative. If they are, then $\mathcal{C} \geq r$. If they are not, then $\mathcal{C} < r$. The bisection algorithm for computing the SSP coefficient of a method is presented in Program 3.2. The following example demonstrates how the conditions are violated if r exceeds the SSP coefficient.

Example 3.6. Consider again the explicit trapezoidal rule method, from Example 2.2. The method has

$$\boldsymbol{\beta}_0 = \begin{pmatrix} 0 & 0 & 0 \\ 1 & 0 & 0 \\ 1/2 & 1/2 & 0 \end{pmatrix}. \tag{3.33}$$

Taking $r = 1/2$, we find

$$\mathbf{v}_{\frac{1}{2}} = \begin{pmatrix} 1 \\ 1/2 \\ 5/8 \end{pmatrix}, \qquad \boldsymbol{\alpha}_{\frac{1}{2}} = \begin{pmatrix} 0 & 0 & 0 \\ 1/2 & 0 & 0 \\ 1/8 & 1/4 & 0 \end{pmatrix}, \tag{3.34}$$

which is the form (2.14), with $\mathcal{C}(\alpha, \beta) = 1/2$. Taking $r = 1$ gives

$$\mathbf{v}_1 = \begin{pmatrix} 1 \\ 0 \\ 1/2 \end{pmatrix}, \qquad \alpha_1 = \begin{pmatrix} 0 & 0 & 0 \\ 1 & 0 & 0 \\ 0 & 1/2 & 0 \end{pmatrix} \tag{3.35}$$

which corresponds to the form (2.15), with $\mathcal{C}(\alpha, \beta) = 1$. On the other hand, if we take $r > 1$, we find that $\alpha_{31} < 0$, so that $\mathcal{C}(\alpha, \beta) = 0$. $\quad\square$

Program 3.2 (Compute the SSP coefficient by bisection).
Evaluates the radius of absolute monotonicity of a Runge–Kutta method, given the Butcher array.
Inputs: Butcher array coefficients in the form of an $m \times m$ matrix **A** *and a column vector* **b** *of length m.*
Output: SSP coefficient C

```
function C = am_radius(A,b,m)
rmax=1000; % rmax >> SSP coefficient of the method
eps=1.e-12; % accuracy tolerance
e=ones(m,1);

beta_0=[A;b'];
rlo=0; rhi=rmax;

while rhi-rlo>eps
   r=0.5*(rhi+rlo);
   [alpha_r,beta_r] = canonical_shu_osher(alpha,beta,m,r);
   v_r = (eye(m+1)-alpha_r)*e;
   if min(min(alpha_r(:)),min(v_r(:)))>=0
      rlo=r;
   else
      rhi=r;
   end
end

if rhi==rmax % r>=rmax
   error('Error: increase value of rmax');
else
   C=rlo;
end
```

3.4   Formulating the optimization problem

Extensive efforts have been made to find optimal explicit SSP Runge–Kutta methods both by analysis and numerical search [62, 26, 27, 97, 98, 87]. Except for [62], all of these efforts formulated the optimization problem using the Shu-Osher form. While this allows the inequality constraints to be written as linear constraints, it leads to a large number of decision variables. By using the conditions of the previous section, the problem can be formulated in terms of the Butcher array only, reducing the number of variables by half and simplifying dramatically the form of the order conditions. We adopt this latter formulation, which can be applied to implicit methods as well:

> **Optimization of SSP coefficient (for given order and number of stages)**
>
> maximize r subject to
> $$\beta_0(I + r\beta_0)^{-1} \geq 0$$
> $$\|r\beta_0(I + r\beta_0)^{-1}\|_\infty \leq 1$$
> $$\tau_k(\beta_0) = 0 \qquad (k \leq p)$$
>
> where τ_k represents the set of order conditions for order k.
> Additional constraints can be added to investigate various subclasses of methods.

This formulation, implemented in MATLAB using a sequential quadratic programming approach (fmincon in the optimization toolbox), was used to find many optimal methods with large number of stages. Some of these methods appear in Chapters 6 and 7.

Remark 3.1. The optimization problem can be reformulated (using a standard approach for converting rational constraints to polynomial constraints) as

$$\max_{\beta_0,\alpha} r \qquad (3.36a)$$

$$\text{subject to} \quad \begin{cases} \alpha \geq 0 \\ \|\alpha\|_\infty \leq 1 \\ r\beta_0 = \alpha(I + r\beta_0), \\ \tau_k(\beta_0) = 0 \qquad (k \leq p) \end{cases} \qquad (3.36b)$$

This optimization problem has only polynomial constraints and thus is appropriate for the BARON optimization software which requires such constraints to be able to guarantee global optimality [88].

$\square$

3.5 Necessity of the SSP time step restriction

In this section we discuss the necessity of the time step restriction (3.11) for strong stability preservation. We show that for any Runge–Kutta method it is possible to find an initial value problem for which monotonicity is satisfied under a given time step restriction for the forward Euler method, but the Runge–Kutta solution violates monotonicity whenever the SSP time step restriction is exceeded.

Theorem 3.3. *Given any irreducible method with SSP coefficient C and a time step $\Delta t > C\Delta t_{FE}$, there exists an initial value problem (2.2) such that the forward Euler condition (2.6) holds with respect to the maximum norm, but monotonicity in this norm is violated by the Runge–Kutta solution.*

Proof. The following constructive proof originally appeared in [57]. It was inspired by the proof of Theorem 2.4 in [96]. Compare also Theorem 5.4 in [62], Theorem 8 in [42], and Theorem 3.4 in [21].

For now, assume that $\mathbf{I}+r\boldsymbol{\beta}_0$ is invertible and let $r > C$. Let $\Delta t = r\Delta t_{FE}$ and let $\boldsymbol{\alpha}, \mathbf{v}$ be defined by (3.24a). Then (3.32) implies $\|[\mathbf{v}\ \ \boldsymbol{\alpha}]\|_\infty > 1$. The idea of the proof is to construct an initial value problem with $\|\mathbf{u}_0\|_\infty = 1$ such that $\|\mathbf{y}_j\|_\infty \geq \|[\mathbf{v}\ \ \boldsymbol{\alpha}]\|_\infty$.

Define $\widehat{\boldsymbol{\alpha}}, \widehat{\mathbf{v}}$ by $\hat{\alpha}_{ij} = \mathrm{sgn}(\alpha_{ij}), \hat{v}_j = \mathrm{sgn}(v_j)$, and let $\hat{\boldsymbol{\alpha}}_j$ denote the jth column of $\widehat{\boldsymbol{\alpha}}$. We will construct an initial value problem with $N = m + 1$ equations, such that the Runge–Kutta stages are given by

$$\mathbf{y}_j = v_j\widehat{\mathbf{v}} + \sum_k \alpha_{jk}\hat{\boldsymbol{\alpha}}_k. \tag{3.37}$$

For the moment, assume that the resulting stages $\mathbf{y}_j$ are distinct. Then we can take

$$\mathbf{u}_0 = \widehat{\mathbf{v}}, \qquad F(\mathbf{u}, t) = \begin{cases} \frac{1}{\Delta t_{FE}}(\hat{\boldsymbol{\alpha}}_j - \mathbf{y}_j) & \text{if } \mathbf{u} = \mathbf{y}_j \text{ for some } j \\ 0 & \text{otherwise.} \end{cases} \tag{3.38}$$

It is straightforward to check that (3.25) is then satisfied, so the $\mathbf{y}_j$ are indeed the stages of the Runge–Kutta solution. The forward Euler condition (2.6) holds, since $F(\mathbf{u}) = 0$ if $\mathbf{u} \neq \mathbf{y}_j$, whereas for $\mathbf{u} = \mathbf{y}_j$ we have

$$\|\mathbf{u}+\Delta t F(\mathbf{u})\|_\infty = \|(1-r)\mathbf{y}_j + r\hat{\boldsymbol{\alpha}}_j\|_\infty \leq (1-r)\|\mathbf{y}_j\| + r\|\hat{\boldsymbol{\alpha}}_j\| \leq \|\mathbf{u}\|_\infty \tag{3.39}$$

since $\|\hat{\boldsymbol{\alpha}}_j\|_\infty \leq 1 \leq \|\mathbf{y}_j\|_\infty$.

The key property of this construction is that

$$\|\mathbf{y}_j\|_\infty \geq (\mathbf{y}_j)_j = v_j \operatorname{sgn}(v_j) + \sum_k \alpha_{jk} \operatorname{sgn}(\alpha_{jk}) = |v_j| + \sum_k |\alpha_{jk}|. \quad (3.40)$$

Hence

$$\max_j \|\mathbf{y}_j\|_\infty \geq \|[\mathbf{v} \quad \boldsymbol{\alpha}]\|_\infty > 1.$$

Thus monotonicity is violated by one of the stages of the method.

This is essentially the construction used in [96] to prove the necessity of the time step restriction (3.11). If $\alpha_{m+1,j} \geq 0$ for all j, then for this example the monotonicity condition (2.7) is still satisfied. However, the example can be modified so that the monotonicity condition is violated by the solution itself (rather than just by the Runge–Kutta stage(s)). To do so, we assume that the method is irreducible, so that every stage influences stage $\mathbf{y}_{m+1}$. Then choose some j such that $\|\mathbf{y}_j\|_\infty > 1$. Replace the jth column of $\hat{\boldsymbol{\alpha}}$ by $\hat{\alpha}_{ij} = \|\mathbf{y}_j\|_\infty \operatorname{sgn}(\alpha_{ij})$. Then (3.37) and (3.38) are still consistent, and the forward Euler condition (2.6) still holds by (3.39). But now for any stage $\mathbf{y}_k$ that is directly influenced by $\mathbf{y}_j$, we have (note the strict inequality, in place of (3.40))

$$\|\mathbf{y}_k\|_\infty \geq (\mathbf{y}_k)_k = |v_k| + \sum_i |\alpha_{ki}| \geq 1. \quad (3.41)$$

Hence, we can modify the kth column of $\hat{\boldsymbol{\alpha}}$ by multiplying it by $\|\mathbf{y}_k\|_\infty$. Then every stage influenced directly by $\mathbf{y}_k$ has maximum norm greater than 1. Proceeding in this manner, since the method is irreducible, we eventually obtain $\|\mathbf{u}^n\|_\infty > 1$.

For the special cases in which $\mathbf{I} + r\boldsymbol{\beta}_0$ is singular or $\mathbf{y}_i = \mathbf{y}_j$ for some $i \neq j$, see pp. 1242-44 of [96]. $\qquad \square$

Example 3.7. Consider the classical four-stage, fourth order Runge–Kutta method, which has

$$\mathbf{A} = \begin{pmatrix} 0 & & & \\ \frac{1}{2} & 0 & & \\ 0 & \frac{1}{2} & 0 & \\ 0 & 0 & 1 & 0 \end{pmatrix}, \quad \mathbf{b} = \left(\frac{1}{6}, \frac{1}{3}, \frac{1}{3}, \frac{1}{6} \right)^{\mathrm{T}},$$

and $\mathcal{C}(\boldsymbol{\beta}_0) = 0$. Taking $r = 1$, we find

$$\mathbf{v}_1 = \left(1, \frac{1}{2}, \frac{3}{4}, \frac{1}{4}, \frac{3}{8} \right)^{\mathrm{T}}, \quad \boldsymbol{\alpha}_1 = \begin{pmatrix} 0 & & & & \\ \frac{1}{2} & 0 & & & \\ -\frac{1}{4} & \frac{1}{2} & 0 & & \\ \frac{1}{4} & -\frac{1}{2} & 1 & 0 & \\ \frac{1}{24} & \frac{1}{4} & \frac{1}{6} & \frac{1}{6} & 0 \end{pmatrix}.$$

So we can construct an initial value problem for which the forward Euler condition is satisfied but max-norm monotonicity is violated by using (3.38), with

$$\widehat{\mathbf{v}} = (1,1,1,1,1)^{\mathrm{T}}, \qquad \widehat{\boldsymbol{\alpha}} = \begin{pmatrix} 0 & & & & \\ 1 & 0 & & & \\ -1 & 1 & 0 & & \\ 1 & -1 & 1 & 0 & \\ 1 & 1 & 1 & 1 & 0 \end{pmatrix}.$$

Computing the Runge–Kutta stages, we find $(y_3)_3 = \frac{3}{4}\hat{v}_3 - \frac{1}{4}\hat{\alpha}_{31} + \frac{1}{2}\hat{\alpha}_{32} = 3/2$ and we have $\|y_3\|_\infty = 3/2$. So we set instead

$$\widehat{\mathbf{v}} = (1,1,1,1,1)^{\mathrm{T}}, \qquad \widehat{\boldsymbol{\alpha}} = \begin{pmatrix} 0 & & & & \\ 1 & 0 & & & \\ -1 & 1 & 0 & & \\ 1 & -1 & 3/2 & 0 & \\ 1 & 1 & 3/2 & 1 & 0 \end{pmatrix}.$$

Then we find

$$(\mathbf{u}^1)_5 = (\mathbf{y}_5)_5 = \frac{3}{8}\hat{v}_5 + \frac{1}{24}\hat{\alpha}_{51} + \frac{1}{4}\hat{\alpha}_{52} + \frac{1}{6}\hat{\alpha}_{53} + \frac{1}{6}\hat{\alpha}_{54} = \frac{13}{12}$$

so $\|\mathbf{u}^1\|_\infty \geq 13/12 > 1 = \|\mathbf{u}_0\|_\infty$. $\square$

Chapter 4

SSP Runge–Kutta Methods for Linear Constant Coefficient Problems

The SSP property is an extraordinarily strong guarantee that holds for any nonlinear initial value problem and convex functional (given the forward Euler condition). Typically, this leads to severe restrictions on the allowable time step and on the possible order of the method. The time step bounds and order barriers on SSP Runge–Kutta methods stem in part from the nonlinearity of the ODEs. The conditions for a method to preserve strong stability in the special case of linear autonomous ODE systems are less restrictive than those required for the nonlinear SSP property. If the ODE is linear, we can obtain a larger allowable time step and higher order methods. In this chapter, we will consider the weaker property of strong stability preservation for the linear, constant coefficient initial value problem

$$\mathbf{u}_t = \mathbf{L}\mathbf{u} \qquad\qquad \mathbf{u}(0) = \mathbf{u}_0, \qquad\qquad (4.1)$$

where $\mathbf{L}$ is a fixed matrix, and we use the boldface letter $\mathbf{u}$ to emphasize that the solution is a vector.

While SSP methods were developed for nonlinear hyperbolic PDEs, a strong stability preserving property is sometimes useful for linear problems as well, for solving linear wave equations such as Maxwell's equations and linear elasticity. Additionally, SSP Runge–Kutta methods for linear problems can be useful from the point of view of stability analysis. For example, in [69], the authors used the energy method to analyze the stability of Runge–Kutta methods for ODEs resulting from coercive approximations such as those in [25]. Using this method it can be proven, for example, that the fourth order Runge–Kutta method preserved a certain stability property with a CFL number of $\frac{1}{31}$. However, using SSP theory for the linear case, we can devise SSP methods of any order with an SSP coefficient of one, and so easily show that the same stability property is preserved in the linear case under a CFL number as large as 1. Finally, the bounds on the

43

size of the SSP coefficient for the linear case are certainly less restrictive than those for the general nonlinear case. Therefore, the possible SSP coefficient for the linear case provides an upper bound for the size of the SSP coefficient in general.

In the following, we present the theory of absolute monotonicity for linear ODEs of the form (4.1). We present the bounds on the obtainable SSP coefficient from the points of view of SSP analysis and of contractivity theory, and show how to derive SSP Runge–Kutta methods of arbitrarily high order for linear problems. Finally, we extend these linear methods to problems which have a time-dependent forcing term or time-dependent boundary conditions.

4.1 The circle condition

In the context of the linear problem (4.1), the forward Euler solution is simply

$$\mathbf{u}^n + \Delta t F(\mathbf{u}^n) = \mathbf{u}^n + \Delta t \mathbf{L}\mathbf{u}^n = (\mathbf{I} + \Delta t \mathbf{L})\mathbf{u}^n.$$

Thus the forward Euler condition $\|u^n + \Delta t \mathbf{L}u^n\| \leq \|u^n\|$ is equivalent to the condition

$$\|\mathbf{I} + \Delta t \mathbf{L}\| \leq 1 \text{ for } 0 < \Delta t \leq \Delta t_{\mathrm{FE}}. \tag{4.2}$$

This is called a *circle condition*, because if $\mathbf{L}$ is normal and $\|\cdot\|$ is the Euclidean norm, then (4.2) simply means that the eigenvalues of $\mathbf{L}$ lie inside a circle in the complex plane centered at $(-1/\Delta t, 0)$ with radius $1/\Delta t$.

Semi-discretizations of hyperbolic PDEs often lead to non-normal matrices and more general convex functionals. For instance, a first order upwind finite difference approximation of the advection equation

$$u_t + u_x = 0 \tag{4.3}$$

takes the form (4.1) with

$$\mathbf{L} = \frac{1}{\Delta x} \begin{pmatrix} -1 & & & \\ 1 & -1 & & \\ & \ddots & \ddots & \\ & & 1 & -1 \end{pmatrix}. \tag{4.4}$$

The exact solution $u(x, t)$ of (4.3), as well as the exact solution of the semi-discrete system given by (4.1) and (4.4), is monotonic (in time) in the maximum norm.

The forward Euler (or circle) condition (4.2) for the semi-discrete system reads

$$\|\mathbf{I} + \Delta t\mathbf{L}\|_\infty = \left\| \begin{pmatrix} 1 - \frac{\Delta t}{\Delta x} & & & \\ \frac{\Delta t}{\Delta x} & 1 - \frac{\Delta t}{\Delta x} & & \\ & \ddots & \ddots & \\ & & \frac{\Delta t}{\Delta x} & 1 - \frac{\Delta t}{\Delta x} \end{pmatrix} \right\|_\infty$$

$$= \left|1 - \frac{\Delta t}{\Delta x}\right| + \left|\frac{\Delta t}{\Delta x}\right| \leq 1.$$

The last inequality holds if and only if $0 \leq \Delta t \leq \Delta x$, so the circle condition holds with $\Delta t_{\text{FE}} = \Delta x$ for this system. As we will see in Section 4.6, this particular system of ODEs holds a special place in the theory of linear strong stability preservation.

4.2   An example: the midpoint method

A common second order Runge–Kutta method is the midpoint method:

$$\mathbf{u}^{(1)} = \mathbf{u}^n + \frac{1}{2}\Delta t F(\mathbf{u}^n)$$

$$\mathbf{u}^{n+1} = \mathbf{u}^n + \Delta t F(\mathbf{u}^{(1)}).$$

It is clear from inspection that the SSP coefficient of this method is 0. Thus we cannot expect to preserve strong stability properties of the solution of an arbitrary (nonlinear) ODE with this method. In fact, the proof of Theorem 3.3 shows how to construct a *nonlinear* ODE such that the forward Euler condition is satisfied under a finite positive time step but the midpoint rule violates monotonicity for any positive time step. However, for any *linear* ODE satisfying the circle condition (4.2), the midpoint method is strong stability preserving under the maximal time step $\Delta t \leq \Delta t_{\text{FE}}$.

To see this, we apply the midpoint method to the linear system (4.1), to obtain

$$\mathbf{u}^{(1)} = \mathbf{u}^n + \frac{1}{2}\Delta t\mathbf{L}\mathbf{u}^n = \left(\mathbf{I} + \frac{1}{2}\Delta t\mathbf{L}\right)\mathbf{u}^n$$

$$\mathbf{u}^{n+1} = \mathbf{u}^n + \Delta t\mathbf{L}\mathbf{u}^{(1)} = \left(\mathbf{I} + \Delta t\mathbf{L} + \frac{1}{2}\Delta t^2\mathbf{L}^2\right)\mathbf{u}^n.$$

Taking norms, we have

$$\|\mathbf{u}^{n+1}\| \leq \left\|\mathbf{I} + \Delta t\mathbf{L} + \frac{1}{2}\Delta t^2\mathbf{L}^2\right\| \cdot \|\mathbf{u}^n\|.$$

Hence strong stability preservation depends on bounding the first factor on the right-hand side. In order to do so, we note that this factor can be rewritten in terms of powers of the factor appearing in the circle condition (4.2):

$$\mathbf{I} + \Delta t \mathbf{L} + \frac{1}{2} \Delta t^2 \mathbf{L}^2 = \frac{1}{2} \mathbf{I} + \frac{1}{2} (\mathbf{I} + \Delta t \mathbf{L})^2.$$

Thus, if the time step satisfies $\Delta t \leq \Delta t_{\mathrm{FE}}$, where Δt_{FE} corresponds to the time step limit in (4.2), we have

$$\begin{aligned}
\|\mathbf{u}^{n+1}\| &\leq \left\| \frac{1}{2}\mathbf{I} + \frac{1}{2}(\mathbf{I} + \Delta t \mathbf{L})^2 \right\| \cdot \|\mathbf{u}^n\| \\
&\leq \frac{1}{2} + \frac{1}{2} \|\mathbf{I} + \Delta t \mathbf{L}\|^2 \cdot \|\mathbf{u}^n\| \\
&\leq \|\mathbf{u}^n\|.
\end{aligned}$$

Hence we see that, although the midpoint method is not strong stability preserving for nonlinear ODEs, it will preserve strong stability properties for linear ODEs provided $\Delta t \leq \Delta t_{\mathrm{FE}}$.

4.3   The stability function

In the example above, we saw that the Runge–Kutta solution of (4.1) was given by a recurrence of the form

$$\mathbf{u}^{n+1} = \psi(\Delta t \mathbf{L})\mathbf{u}^n, \tag{4.5}$$

where $\psi(z) = 1 + z + \frac{1}{2}z^2$. In fact, application of any Runge–Kutta method to (4.1) will yield a solution given by (4.5), where the function $\psi(z)$ depends on the particular Runge–Kutta method.

Thus the behavior of the numerical solution is determined entirely by $\psi(z)$. In particular, for large n the solution will grow without bound if $\|\psi(\Delta t \mathbf{L})\| > 1$, and it will approach zero if $\|\psi(\Delta t \mathbf{L})\| < 1$. For this reason we refer to ψ as the stability function of the Runge–Kutta method.

4.3.1   *Formulas for the stability function*

In order to determine $\psi(z)$ in general, we will again employ the compact notation of the last chapter, and be careful to use the dimensionally correct $\bar{A} = \mathbf{I} \otimes \mathbf{A}$ and $\bar{b} = \mathbf{I} \otimes \mathbf{b}$ where appropriate.

Thus we write the Runge–Kutta solution of (4.1) as

$$\boldsymbol{y} = \boldsymbol{e}u^n + z\bar{\boldsymbol{A}}\boldsymbol{u} \qquad (4.6a)$$

$$\mathbf{u}^{n+1} = \mathbf{u}^n + z\bar{\boldsymbol{b}}^{\mathrm{T}}\boldsymbol{u}. \qquad (4.6b)$$

We can rewrite (4.6a) as

$$(\boldsymbol{I} - \bar{\boldsymbol{A}}z)\boldsymbol{y} = \boldsymbol{e}u^n$$

$$\boldsymbol{y} = (\boldsymbol{I} - \bar{\boldsymbol{A}}z)^{-1}\boldsymbol{e}u^n. \qquad (4.7)$$

Combining (4.7) with (4.6b) gives

$$\mathbf{u}^{n+1} = \left(1 + z\bar{\boldsymbol{b}}^{\mathrm{T}}(\boldsymbol{I} - \bar{\boldsymbol{A}}z)^{-1}e\right)\mathbf{u}^n,$$

so

$$\psi(z) = 1 + z\mathbf{b}^{\mathrm{T}}(\mathbf{I} - \mathbf{A}z)^{-1}\mathbf{e}. \qquad (4.8)$$

Example 4.3.1. We compute ψ based on this formula, using the midpoint method as an example again. The Butcher representation of the method is

$$\mathbf{A} = \begin{pmatrix} 0 & 0 \\ \frac{1}{2} & 0 \end{pmatrix} \qquad \mathbf{b} = \begin{pmatrix} 0 \\ 1 \end{pmatrix}.$$

So

$$(\mathbf{I} - \mathbf{A}z)^{-1} = \begin{pmatrix} 1 & 0 \\ -\frac{z}{2} & 1 \end{pmatrix}^{-1} = \begin{pmatrix} 1 & 0 \\ \frac{z}{2} & 1 \end{pmatrix}.$$

Thus

$$\psi(z) = 1 + z\begin{pmatrix} 0 & 1 \end{pmatrix}\begin{pmatrix} 1 & 0 \\ \frac{z}{2} & 1 \end{pmatrix}\begin{pmatrix} 1 \\ 1 \end{pmatrix}$$

$$= 1 + z\begin{pmatrix} 0 & 1 \end{pmatrix}\begin{pmatrix} 1 \\ \frac{z}{2} + 1 \end{pmatrix}$$

$$= 1 + z + \frac{1}{2}z^2.$$

An alternative formula for the stability function can be obtained using Cramer's rule. If we define again

$$\boldsymbol{\beta}_0 = \begin{pmatrix} \mathbf{A} & 0 \\ \mathbf{b}^{\mathrm{T}} & 0 \end{pmatrix} \qquad (4.9a)$$

then we can write (4.6a) and (4.6b) as a single linear system:

$$(\mathbf{I} - z\boldsymbol{\beta}_0)\boldsymbol{y} = \boldsymbol{e}u^n.$$

Writing out this system of equations, we have

$$
\begin{pmatrix}
1 - za_{11} & -za_{12} & \cdots & -za_{1m} & 0 \\
-za_{21} & 1 - za_{22} & \cdots & -za_{2m} & 0 \\
\vdots & \vdots & \ddots & \vdots & \vdots \\
-za_{m1} & -za_{m2} & \cdots & 1 - za_{mm} & 0 \\
-zb_1 & -zb_2 & \cdots & -zb_m & 1
\end{pmatrix}
\begin{pmatrix}
\mathbf{u}^{(1)} \\
\mathbf{u}^{(2)} \\
\vdots \\
\mathbf{u}^{(m)} \\
\mathbf{u}^{n+1}
\end{pmatrix}
=
\begin{pmatrix}
\mathbf{u}^n \\
\mathbf{u}^n \\
\vdots \\
\mathbf{u}^n \\
\mathbf{u}^n
\end{pmatrix} .
\tag{4.10}
$$

Cramer's rule says that $\mathbf{u}^{n+1}$ is given by the ratio of two determinants. The denominator is $\det(\mathbf{I} - z\boldsymbol{\beta}_0)$, which is equal to $\det(\mathbf{I} - z\mathbf{A})$ because the last column of $\boldsymbol{\beta}_0$ is zero. The numerator is the matrix obtained by replacing the last column of $\mathbf{I} - z\boldsymbol{\beta}_0$ with the right-hand side of (4.10):

$$
\begin{pmatrix}
1 - za_{11} & -za_{12} & \cdots & -za_{1m} & \mathbf{u}^n \\
-za_{21} & 1 - za_{22} & \cdots & -za_{2m} & \mathbf{u}^n \\
\vdots & \vdots & \ddots & \vdots & \vdots \\
-za_{m1} & -za_{m2} & \cdots & 1 - za_{mm} & \mathbf{u}^n \\
-zb_1 & -zb_2 & \cdots & -zb_m & \mathbf{u}^n
\end{pmatrix} .
\tag{4.11}
$$

Subtracting the last row of (4.11) from the first m rows gives

$$
\begin{pmatrix}
1 - za_{11} + zb_1 & -za_{12} + zb_2 & \cdots & -za_{1m} + zb_m & 0 \\
-za_{21} + zb_1 & 1 - za_{22} + zb_2 & \cdots & -za_{2m} + zb_m & 0 \\
\vdots & \vdots & \ddots & \vdots & \vdots \\
-za_{m1} + zb_1 & -za_{m2} + zb_2 & \cdots & 1 - za_{mm} + zb_m & 0 \\
-zb_1 & -zb_2 & \cdots & -zb_m & \mathbf{u}^n
\end{pmatrix} .
\tag{4.12}
$$

Due to the nature of the last column of (4.12), we see that the determinant of this matrix is equal to $\mathbf{u}^n \det(\mathbf{I} - z\mathbf{A} + z\mathbf{e}\mathbf{b}^{\mathrm{T}})$. Thus we have the formula

$$
\psi(z) = \frac{\det(\mathbf{I} - z\mathbf{A} + z\mathbf{e}\mathbf{b}^{\mathrm{T}})}{\det(\mathbf{I} - z\mathbf{A})} .
$$

The Butcher array formulation thus gives rise to two forms of the stability function:

$$
\psi(z) = 1 + z\mathbf{b}^{\mathrm{T}}(\mathbf{I} - \mathbf{A}z)^{-1}\mathbf{e}
$$

and

$$
\psi(z) = \frac{\det(\mathbf{I} - z\mathbf{A} + z\mathbf{e}\mathbf{b}^{\mathrm{T}})}{\det(\mathbf{I} - z\mathbf{A})} .
$$

Similarly, two forms of the stability function can be obtained directly in terms of the Shu-Osher coefficients. We rewrite the modified Shu-Osher form a little differently here:

$$
\boldsymbol{y} = \bar{v}^*\mathbf{u}^n + \bar{\alpha}^*\boldsymbol{y} + z\bar{\beta}^*\boldsymbol{y}
\tag{4.13}
$$

$$
\mathbf{u}^{n+1} = \bar{v}^*_{m+1}\mathbf{u}^n + \bar{\alpha}^*_{m+1}\boldsymbol{y} + z\bar{\beta}^*_{m+1}\boldsymbol{y},
\tag{4.14}
$$

where $\boldsymbol{\alpha}^*$ (or $\boldsymbol{\beta}^*$, $\mathbf{v}^*$) is the matrix made up of the first m rows of $\boldsymbol{\alpha}$ (or $\boldsymbol{\beta}$, $\mathbf{v}$), and $\boldsymbol{\alpha}^*_{m+1}$ ($\boldsymbol{\beta}^*_{m+1}$, v^*_{m+1}) is the vector made up of the last row of $\boldsymbol{\alpha}$ (or $\boldsymbol{\beta}$, $\mathbf{v}$). As before, the bars on these quantities indicate the Kronecker product of the matrix with the identity.

We can obtain the stability function by solving (4.13) for $\boldsymbol{y}$ and substituting into (4.14), to give

$$\mathbf{u}^{n+1} = \left(\bar{v}^*_{m+1} + (\bar{\boldsymbol{\alpha}}^*_{m+1} + z\bar{\boldsymbol{\beta}}^*_{m+1})(\mathbf{I} - \bar{\boldsymbol{\alpha}}^* - z\bar{\boldsymbol{\beta}}^*)^{-1}\bar{v}^* \right) \mathbf{u}^n. \qquad (4.15)$$

Thus

$$\psi(z) = \left(v^*_{m+1} + (\boldsymbol{\alpha}^*_{m+1} + z\boldsymbol{\beta}^*_{m+1})(\mathbf{I} - \boldsymbol{\alpha}^* - z\boldsymbol{\beta}^*)^{-1}\mathbf{v}^* \right). \qquad (4.16)$$

Observe that by taking $\boldsymbol{\alpha} = 0$, (4.16) reduces to (4.8).

The second form follows from Cramer's rule, and can be derived in a manner similar to that above. The result is

$$\psi(z) = \frac{\det(\mathbf{I} - \boldsymbol{\alpha}^* - z\boldsymbol{\beta}^* + \mathbf{v}(\boldsymbol{\alpha}^*_{m+1} + z\boldsymbol{\beta}^*_{m+1}))}{\det(\mathbf{I} - \boldsymbol{\alpha}^* - z\boldsymbol{\beta}^*)}. \qquad (4.17)$$

4.3.2   An alternative form

From any of the four equivalent formulas above, it is apparent that the stability function $\psi(z)$ of an m-stage explicit Runge–Kutta method is a polynomial of degree at most m:

$$\psi(z) = a_0 + a_1 z + a_2 z^2 + \cdots + a_m z^m = \sum_{j=0}^{m} a_j z^j. \qquad (4.18)$$

As we saw in Section 4.2, analysis of the SSP property is more convenient if we use a different representation of ψ:

$$\psi_r(z) = \gamma_0 + \gamma_1 \left(1 + \frac{z}{r}\right) + \gamma_2 \left(1 + \frac{z}{r}\right)^2 + \cdots + \gamma_m \left(1 + \frac{z}{r}\right)^m = \sum_{i=0}^{m} \gamma_i \left(1 + \frac{z}{r}\right)^i. \qquad (4.19)$$

In this second representation, we use the basis functions $\left(1 + \frac{z}{r}\right)^i$ in place of the usual monomial basis. Thus, the coefficients a_i in (4.18) and γ_i in (4.19) are just the coordinates of ψ in the two bases. Expanding terms in (4.19) using the binomial formula and then collecting powers of z gives the following relation between them:

$$a_i = \frac{1}{i! r^i} \sum_{j=0}^{m} \gamma_j \prod_{n=0}^{i-1} (j - n). \qquad (4.20)$$

Example 4.3.2. Solving the system above for the stability function of the midpoint method, we find that it can be written in the form

$$\psi(z) = 1 - r + \frac{1}{2}r^2 + (r - r^2)\left(1 + \frac{z}{r}\right) + \frac{1}{2}r^2\left(1 + \frac{z}{r}\right)^2.$$

Taking $r = 1$ gives the form

$$\psi(z) = \frac{1}{2} + \frac{1}{2}(1 + z)^2$$

that we used to establish the strong stability preserving property at the beginning of the chapter. Notice that if we take $r > 1$, the coefficient of the $(1 + z/r)$ term becomes negative.

4.3.3   *Order conditions on the stability function*

Since the exact solution of (4.1) is given by

$$\mathbf{u}(t + \Delta t) = \exp(\mathbf{L}\Delta t)\mathbf{u}(t) = \sum_{j=0}^{\infty} \frac{1}{j!}(\Delta t\mathbf{L})^j \mathbf{u}(t)$$

we see that the order of accuracy of the numerical solution (4.5) is just given by the order of agreement between the stability function and the exponential function when Δt is small. In other words, the conditions for a method to have order p are

$$a_j = \frac{1}{j!} \text{ for all } j \leq p. \tag{4.21}$$

In particular, any consistent method must have $a_0 = 1$. According to (4.20), $a_0 = \sum_j \gamma_j$, so any consistent method satisfies

$$\sum_j \gamma_j = 1. \tag{4.22}$$

4.4   Strong stability preservation for linear systems

The usefulness of the polynomial representation (4.19) is due to its connection with the circle condition (4.2). The parameter r plays the role of the ratio between the chosen time step and the forward Euler time step:

$$\Delta t = r\Delta t_{\text{FE}}.$$

Replacing z with $\Delta t\mathbf{L}$ in (4.19), we have

$$\mathbf{I} + \frac{z}{r} = \mathbf{I} + \frac{\Delta t\mathbf{L}}{\Delta t/\Delta t_{\text{FE}}} = \mathbf{I} + \Delta t_{\text{FE}}\mathbf{L}.$$

Thus, if the circle condition (4.2) is satisfied, then

$$\left\|\left(\mathbf{I}+\frac{z}{r}\right)^i\right\| = \|(\mathbf{I}+\Delta t_{\text{FE}}\mathbf{L})\|^i \le 1.$$

So if all the coefficients γ_j are non-negative, then (using (4.22))

$$\|\mathbf{u}^n\| = \|\psi(\Delta t\mathbf{L})\mathbf{u}^{n-1}\|$$

$$= \left\|\sum_j \gamma_j \left(\mathbf{I}+\frac{\Delta t}{r}\mathbf{L}\right)^j \mathbf{u}^{n-1}\right\|$$

$$\le \sum_j \gamma_j \|\mathbf{I}+\Delta t_{\text{FE}}\mathbf{L}\|^j \|\mathbf{u}^{n-1}\|$$

$$\le \|\mathbf{u}^{n-1}\| \sum_j \gamma_j = \|\mathbf{u}^{n-1}\|.$$

Thus we see that strong stability preservation for (4.1) depends on choosing a time step so that the ratio $r = \Delta t/\Delta t_{\text{FE}}$ leads to positive coefficients γ_j of the stability polynomial in the representation (4.19). In practice, we wish to use the largest possible time step ratio $r = \Delta t/\Delta t_{\text{FE}}$ such that $\gamma_j \ge 0$.

4.5 Absolute monotonicity

Let us now determine for which values of r we have $\gamma_j \ge 0$. Observe that the form (4.19) is closely related to the Taylor expansion of ψ about $z = -r$:

$$\psi(z) = \sum_{j=0}^{m} \frac{(z+r)^j}{j!}\psi^{(j)}(-r) \tag{4.23}$$

$$= \sum_{j=0}^{m} \frac{r^j}{j!}\left(\frac{z}{r}+1\right)^j \psi^{(j)}(-r). \tag{4.24}$$

Comparing this with (4.19), we see that the coefficients there are given by

$$\gamma_j = \frac{r^j}{j!}\psi^{(j)}(-r). \tag{4.25}$$

Thus γ_j is non-negative if the jth derivative of ψ is non-negative at $z = -r$. This brings us to the concept of absolute monotonicity.

Definition 4.1. Absolute monotonicity of a function. A function $\psi(z) : \mathbb{R} \to \mathbb{R}$ is absolutely monotonic at x if $\psi(z)$ and all of its derivatives exist and are non-negative at $z = x$.

Hence a method is SSP for the linear equation (4.1) if the time step is chosen so that ψ is absolutely monotonic at $-r = -\Delta t/\Delta t_{\mathrm{FE}}$. It would be problematic if choosing a *smaller* time step (i.e. a smaller value of r) could lead to violation of the SSP property. The following lemma, which is analogous to Lemma 3.3, states that this is not possible. The statement holds more generally, but for simplicity we will state and prove it for polynomials only.

Lemma 4.1. *For $\xi > 0$, a polynomial $\psi(z)$ is absolutely monotonic at $z = -\xi$ if and only if it is absolutely monotonic on the interval $z \in (-\xi, 0]$.*

Proof. Suppose $\psi(z)$ is absolutely monotonic on $(-\xi, 0]$. Then by continuity, ψ is absolutely monotonic at $-\xi$.

On the other hand, suppose $\psi(z)$ is absolutely monotonic at $z = -\xi < 0$ and take $0 \leq \eta \leq \xi$. Then writing ψ in form (4.19) with $r = \xi$, using the fact that $|1 + \eta/\xi| \leq 1$, and differentiating term-by-term shows that $\psi(z)$ and its derivatives are non-negative at $z = -\eta$.

Hence we can characterize the absolute monotonicity of a function in terms of the left endpoint of a single interval:

Definition 4.2. Radius of absolute monotonicity. The radius of absolute monotonicity $R(\psi)$ of a function $\psi : \mathbb{R} \to \mathbb{R}$ is the largest value of r such that $\psi(z)$ and all of its derivatives exist and are non-negative for $z \in (-r, 0]$.

The coefficients γ_j in (4.19) are non-negative if and only if $r > 0$ is less than the radius of absolute monotonicity:

$$\gamma_j \geq 0 \iff r \leq R(\psi). \tag{4.26}$$

In other words, strong stability is preserved if the time step ratio is less than the radius of absolute monotonicity:

$$\frac{\Delta t}{\Delta t_{\mathrm{FE}}} \leq R(\psi).$$

We summarize this conclusion in the following definition and theorem.

Definition 4.3. Threshold factor of a Runge–Kutta method. For a given Runge–Kutta method with stability function ψ, we refer to the radius of absolute monotonicity $R = R(\psi)$ as the **threshold factor** of the Runge–Kutta method.

Theorem 4.1. *Let the matrix* $\mathbf{L}$ *and convex functional* $\|\cdot\|$ *be such that the circle condition (4.2) is satisfied. Then the monotonicity property (2.7) holds for the solution of the linear autonomous initial value problem (4.1) by a consistent Runge–Kutta method (2.10) if the time step satisfies*

$$0 \le \Delta t \le R\Delta t_{\mathrm{FE}}, \qquad (4.27)$$

where R *is the threshold factor of the method.*

The essence of the proof is the observation that form (4.19) expresses the method as a convex combination of iterated forward Euler steps, in analogy to the proof of Theorem 2.1. Observe that the time step restriction (4.27), like (2.8), involves two factors: Δt_{FE}, which depends only on $\mathbf{L}$ (i.e. on the particular system of ODEs), and R, which depends only on the numerical method.

4.6 Necessity of the time step condition

We now give an example that demonstrates the sense in which absolute monotonicity of ψ is a *necessary* condition for strong stability preservation for linear systems. Recall that we had an analogous result in Theorem 3.3, but note that the formulation for the linear case is simpler. Consider again the upwind finite difference semi-discretization (4.4) of the advection equation (4.3). Recall that the exact solution is monotonic in the maximum norm $\|\cdot\|_{\infty}$.

We will show that the time step restriction (4.27) is strictly necessary in this case. That is, if a Runge–Kutta method with stability function ψ is applied to this problem and time step restriction (4.27) is violated, then there exists an initial condition $\mathbf{u}_0$ such that $\|\mathbf{u}^1\|_{\infty} > \|\mathbf{u}_0\|_{\infty}$.

Theorem 4.2. *For any function* ψ *and for* $\mathbf{L}$ *given by (4.4), we have*

$$\|\psi(\Delta t\mathbf{L})\|_{\infty} \le 1 \quad \text{iff} \quad \Delta t \le R(\psi)\Delta t_{\mathrm{FE}}.$$

Hence, for a Runge–Kutta method with stability function ψ, *monotonicity in the maximum norm is guaranteed for the solution of the initial value problem (4.1) if and only if the time step satisfies (4.27).*

Proof. The sufficiency of the time step restriction follows from Theorem 4.1. To show that this restriction is necessary, assume $\|\psi(\Delta t\mathbf{L})\|_{\infty} \le 1$. Then taking $\Delta t = r\Delta x$, we have

$$\Delta t\mathbf{L} = \mathbf{Z} = -r\mathbf{I} + r\mathbf{E},$$

where

$$\mathbf{E} = \begin{pmatrix} 0 & & & \\ 1 & 0 & & \\ & \ddots & \ddots & \\ & & 1 & 0 \end{pmatrix}.$$

So, expanding ψ about $-r\mathbf{I}$,

$$\psi(\mathbf{Z}) = \sum_{j=0}^{\infty} \frac{(\mathbf{Z}+r\mathbf{I})^j}{j!} \psi^{(j)}(-r) = \sum_{j=0}^{\infty} \frac{r^j}{j!} \psi^{(j)}(-r)\mathbf{E}^j = \sum_{j=0}^{\infty} \gamma_j \mathbf{E}^j,$$

where γ_j is defined in (4.19). Since

$$\sum_{j=0}^{\infty} \gamma_j \mathbf{E}^j = \begin{pmatrix} \gamma_0 & & & \\ \gamma_1 & \gamma_0 & & \\ \vdots & & \ddots & \ddots \\ \gamma_{N-1} & \cdots & \gamma_1 & \gamma_0 \end{pmatrix},$$

then

$$\sum_{j=0}^{N-1} |\gamma_j| = \|\psi(\mathbf{Z})\|_{\infty} \le 1 = \sum_{j=0}^{\infty} \gamma_j,$$

where the last equality is just the consistency condition (4.22). Since this holds for any positive integer N, we have

$$\sum_{j=0}^{\infty} |\gamma_j| \le \sum_{j=0}^{\infty} \gamma_j,$$

so $\gamma_j \ge 0$. Thus ψ is absolutely monotonic at $-r$, so $\Delta t = r\Delta x \le R(\psi)\Delta t_{\text{FE}}$. $\square$

4.7   An optimization algorithm

A practical question that arises is how to find methods with maximal threshold factor within a given class. We use $R_{m,p}$ to denote the maximal threshold factor among all order p Runge–Kutta methods with up to m stages. The following efficient algorithm for the determination of optimal methods for any m, p was given in [55].

The problem of finding $R_{m,p}$ can be written by combining the order conditions (4.21) and the absolute monotonicity condition:

maximize r subject to

$$a_i = \frac{1}{i!} \qquad\qquad (0 \le i \le p)$$

$$\gamma(z) \ge 0 \qquad\qquad (-r \le z \le 0),$$

where the coefficients $\gamma(z)$ are found by solving the linear system (4.20). This appears to be a difficult problem, because the domain of the inequality constraints depends on the objective function, r. Representing ψ in the form (4.19) leads to the alternate formulation [60]

maximize r subject to

$$\sum_{j=0}^{m} \gamma_j \prod_{n=0}^{i-1} (j - n) = r^i \qquad\qquad 0 \le i \le p \qquad (4.28a)$$

$$\gamma \ge 0. \qquad (4.28b)$$

This is still a nonlinear problem, but *for any fixed value of* r it is simply a linear programming feasibility problem. According to Lemma 4.1, this problem will be feasible if $R_{m,p} \ge r \ge 0$. Hence we can use bisection to find $R_{m,p}$.

Algorithm 4.1 (Compute the threshold factor).
Computes by bisection the threshold factor R of a polynomial of degree m that matches the p order conditions.
Inputs: Positive integers m, p and real numbers ϵ and r_{max} such that $R_{m,p} \le r_{max}$
Output: r satisfying $R_{m,p} - \epsilon \le r \le R_{m,p}$
1. $r_{min} := 0$.
2. Set $r := (r_{max} + r_{min})/2$.
3. Determine whether there exists a polynomial of degree at most m that is absolutely monotonic at $-r$. If so, set $r_{min} := r$; otherwise set $r_{max} := r$.
4. If $r_{max} - r_{min} < \epsilon$, set $r := r_{min}$ and stop. Otherwise, return to step 2.

4.8 Optimal SSP Runge–Kutta methods for linear problems

Explicit Runge–Kutta methods with positive threshold factor exist for arbitrarily high orders. Optimally contractive explicit Runge–Kutta methods were studied by Kraaijevanger in [60], where optimal methods were given for many values of m and p, including $1 \leq p \leq m \leq 10$, and $p \in \{1, 2, 3, 4, m - 1, m - 2, m - 3, m - 4\}$ for any m. It has been shown that an m-stage Runge–Kutta method of order p for linear problems can have threshold factor no greater than $m - p + 1$. Below, we give the families of SSP Runge–Kutta methods for linear problems that achieve the bound $R_{m,p} = m - p + 1$. These methods can be found in [60, 28], and are given below in canonical Shu-Osher form (2.10).

SSPRK linear (m,1): Let us begin with the class of first order, m-stage methods with $R_{m,p} = m$

$$\mathbf{u}^{(0)} = \mathbf{u}^n$$
$$\mathbf{u}^{(i)} = \left(1 + \frac{\Delta t}{m}\mathbf{L}\right)\mathbf{u}^{(i-1)}, i = 1, ..., m \qquad (4.29)$$
$$\mathbf{u}^{n+1} = \mathbf{u}^{(m)}.$$

The SSP coefficient can be easily verified by observing that for each nonzero α_{ij} we have $\alpha_{ij} = 1$ and $\beta_{ij} = \frac{1}{m}$, and $\beta_{ij} = 0$ whenever $\alpha_{ij} = 0$. To check the order, we notice that

$$\mathbf{u}^{n+1} = \left(1 + \Delta t \frac{1}{m}\mathbf{L}\right)^m \mathbf{u}^n$$
$$= \left(1 + \Delta t\mathbf{L} + \Delta t^2 \frac{m-1}{2m}\mathbf{L}^2 + \cdots + \frac{1}{m^m}\Delta t^m \mathbf{L}^m\right)\mathbf{u}^n$$

since $\frac{m-1}{2m} \neq \frac{1}{2}$ for any m, we have

$$\mathbf{u}^{n+1} = \left(1 + \Delta t\mathbf{L} + O(\Delta t^2)\right)\mathbf{u}^n,$$

which is a first order method. Although this method has a significantly higher threshold factor than the forward Euler method, it has a correspondingly higher computational cost. The step size can be increased with the number of stages, but then the computational cost is also increased by the same factor. For this reason, we define the *effective* threshold factor $R_{\text{eff}} = \frac{R_{m,p}}{m}$. In this case, we have an effective threshold factor $R_{\text{eff}} = 1$.

SSPRK linear (m,2): The m-stage, second order SSP methods

$$\mathbf{u}^{(0)} = \mathbf{u}^n$$

$$\mathbf{u}^{(i)} = \left(1 + \frac{\Delta t}{m-1}\mathbf{L}\right)\mathbf{u}^{(i-1)}, \qquad i = 1, ..., m-1$$

$$\mathbf{u}^m = \frac{1}{m}\mathbf{u}^{(0)} + \frac{m-1}{m}\left(1 + \frac{\Delta t}{m-1}\mathbf{L}\right)\mathbf{u}^{(m-1)}$$

$$\mathbf{u}^{n+1} = \mathbf{u}^{(m)}$$

have threshold factor $R = m - 1$, and effective threshold factor $R_{\text{eff}} = \frac{m-1}{m}$. Although these methods were designed for linear problems, this class of methods is also second order and SSP for nonlinear problems, and is the same family of methods mentioned in Section 6.2.1.

SSPRK linear (m,m−1): The m-stage, order $p = m - 1$ method is given by

$$\mathbf{u}^{(0)} = \mathbf{u}^n$$

$$\mathbf{u}^{(i)} = \mathbf{u}^{(i-1)} + \frac{1}{2}\Delta t\mathbf{L}\mathbf{u}^{(i-1)}, \qquad i = 1, ..., m-1$$

$$\mathbf{u}^{(m)} = \sum_{k=0}^{m-2} \alpha_k^m \mathbf{u}^{(k)} + \alpha_{m-1}^m\left(\mathbf{u}^{(m-1)} + \frac{1}{2}\Delta t\mathbf{L}\mathbf{u}^{(m-1)}\right),$$

$$\mathbf{u}^{n+1} = \mathbf{u}^{(m)}.$$

Where the coefficients α_k^m of the final stage of the m-stage method are given iteratively by

$$\alpha_k^m = \frac{2}{k}\alpha_{k-1}^{m-1}, \qquad k = 1, ..., m-2$$

$$\alpha_{m-1}^m = \frac{2}{m}\alpha_{m-2}^{m-1}, \qquad \alpha_0^m = 1 - \sum_{k=1}^{m-1} \alpha_k^m,$$

starting from the coefficients of the 2-stage, first order method $\alpha_0^2 = 0$ and $\alpha_1^2 = 1$. These methods are SSP with optimal threshold factor $R = 2$, and effective threshold factor $R_{\text{eff}} = \frac{2}{m}$.

SSPRK linear (m,m): The m-stage mth order schemes given by

$$\mathbf{u}^{(0)} = \mathbf{u}^n$$

$$\mathbf{u}^{(i)} = \mathbf{u}^{(i-1)} + \Delta t\mathbf{L}\mathbf{u}^{(i-1)}, \qquad i = 1, ..., m-1$$

$$\mathbf{u}^{(m)} = \sum_{k=0}^{m-2} \alpha_k^m \mathbf{u}^{(k)} + \alpha_{m-1}^m\left(\mathbf{u}^{(m-1)} + \Delta t\mathbf{L}\mathbf{u}^{(m-1)}\right),$$

$$\mathbf{u}^{n+1} = \mathbf{u}^{(m)}.$$

Where the coefficients α_k^m of the final stage of the m-stage method are given iteratively by

$$\alpha_k^m = \frac{1}{k}\alpha_{k-1}^{m-1}, \qquad k = 1, ..., m-2$$

$$\alpha_{m-1}^m = \frac{1}{m!}, \qquad \alpha_0^m = 1 - \sum_{k=1}^{m-1} \alpha_k^m,$$

starting from the coefficient of the forward Euler method $\alpha_0^1 = 1$. These methods are SSP with optimal threshold factor $R = 1$, and the effective threshold factor $R_{\text{eff}} = \frac{1}{m}$.

4.9 Linear constant coefficient operators with time dependent forcing terms

In practice, we often have linear autonomous equations with time-dependent boundary conditions or a time dependent forcing term. These equations can be written

$$\boldsymbol{u}_t = \mathbf{L}\boldsymbol{u} + \boldsymbol{f}(t). \tag{4.30}$$

Such equations arise, for example, from a linear PDE with time dependent boundary conditions such as Maxwell's equations (see [11]). We would like to extend the SSP methods for autonomous ODEs listed in the section above to the case of a constant linear operator with a time dependent forcing term. This ODE is a linear time dependent ODE and as such, the Runge–Kutta methods derived above for a linear time-invariant ODE will not have the correct order. The problem is that the class of RK methods for linear, time dependent ODEs is not equivalent to those for linear time invariant ODEs [107]. However, if the functions $f(t)$ can be written in a suitable way, then we can convert Equation (4.30) to a linear constant-coefficient ODE.

To see how this can be done, we recall that the order conditions for a Runge–Kutta method are derived for an autonomous system $y'(x) = g(y(x))$ but can then be applied to non-autonomous system. The reason for this is that any non-autonomous system of the form

$$u'(x) = h(x, u(x))$$

can be converted to an autonomous system by setting

$$\boldsymbol{y}(x) = \begin{pmatrix} x \\ u(x) \end{pmatrix}$$

and then

$$y'(x) = \begin{pmatrix} 1 \\ u'(x) \end{pmatrix} = \begin{pmatrix} 1 \\ h(x, u) \end{pmatrix} = g(y).$$

This does not help us directly, though, because a linear non-autonomous equation can become nonlinear when converted to an autonomous equation. We can, however convert the linear autonomous equation to a linear non-autonomous equation if we are willing to *approximate* $f(t)$ as a linear combination of basis functions which will then be related by a linear autonomous differential equation to their derivatives.

Starting with the system (4.30), we approximate each element of the vector $f(t)$ by $f_i(t) = \sum_{j=0}^{n} g_{ij} q_j(t)$ so that

$$f(t) = \mathbf{G} q(t)$$

where $\mathbf{G} = [G_{i,j}] = [g_{ij}]$ is a constant matrix and $q(t) = [q_j(t)]$ are a set of functions which have the property that $q'(t) = Dq(t)$, where $\mathbf{D}$ is a constant matrix. Now we have a linear, constant coefficient ODE

$$q'(t) = \mathbf{D} q(t)$$
$$u'(t) = \mathbf{L} u(t) + \mathbf{G} q(t),$$

which can be written

$$y_t = \mathbf{M} y(t) \tag{4.31}$$

where

$$y(t) = \begin{pmatrix} q(t) \\ u(t) \end{pmatrix}$$

and

$$\mathbf{M} = \begin{pmatrix} \mathbf{D} \ 0 \\ \mathbf{G} \ \mathbf{L} \end{pmatrix}.$$

Thus, an equation of the form (4.30) can be approximated (or given exactly) by a linear constant coefficient ODE, and the SSP Runge–Kutta methods for linear constant coefficient operators can be applied to guarantee that any strong stability properties satisfied with forward Euler will be preserved. However, it is important to check that the forward Euler condition is still satisfied for the modified equation (4.31).

Remark 4.1. There are many possible ways to approximate the functions $f(t)$. One simple approach is to use the polynomials $q_j(t) = t^j$. In this case, the differentiation matrix $\mathbf{D}$ is given by

$$D_{ij} = \begin{cases} i - 1 & \text{if } j = i - 1 \\ 0 & \text{otherwise.} \end{cases}$$

A better approximation can be obtained using the Chebyshev polynomials. For these polynomials the relationship between the polynomial and its derivative is given by $T'_n(t) = \sum_{j=0}^{\infty} b_j T_j(t)$ where

$$b_j = \begin{cases} n & \text{for } j = 0, \text{ if } n \text{ is odd} \\ 2n & \text{for } j > 0, \text{ if } j + n \text{ is odd.} \end{cases}$$

This form allows an easy computation of the differentiation matrix. Alternatively, one could use a Fourier approximation of the function for which the differentiation matrix has a simple form. In general, it is best to allow the form of the function $f(t)$ to determine the approximation whenever possible. Furthermore, the Runge–Kutta method used must be of high enough order to evolve the ODE for the function $f(t)$ to a high enough resolution. $\qquad\square$

Example 4.1. Given the problem

$$u_t = u_{xx} + 4t^3 \qquad 0 \le x \le \pi$$

with initial condition $u(x, 0) = sin(x)$ and boundary conditions $u(0, t) = u(\pi, t) = t^4$, we employ the second order centered difference spatial discretization for N points

$$u_{xx} \approx \frac{u_{j+1} - 2u_j + u_{j-1}}{\Delta x^2},$$

which gives us the $N \times N$ linear operator

$$\mathbf{L}_{ij} = \begin{cases} \frac{1}{\Delta x^2} & \text{if } j = i - 1 \text{ or } j = i + 1 \\ -\frac{2}{\Delta x^2} & \text{if } j = i \\ 0 & \text{otherwise} \end{cases}$$

in

$$\boldsymbol{u}_t = \mathbf{L}\boldsymbol{u} + 4t^3 \boldsymbol{e}. \tag{4.32}$$

To incorporate the time-dependent boundary conditions as well as the time dependent forcing term, we define

$$\boldsymbol{q}^{\mathrm{T}} = \left(1, t, t^2, t^3, t^4\right)$$

and

$$\boldsymbol{y}^{\mathrm{T}} = \left(\boldsymbol{q}^{\mathrm{T}}, \boldsymbol{u}^{\mathrm{T}}\right)$$

so the ODE becomes

$$\boldsymbol{y}_t = \begin{pmatrix} \mathbf{D} \ 0 \\ \mathbf{G} \ \mathbf{L} \end{pmatrix} \boldsymbol{y},$$

where

$$D_{ij} = \begin{cases} i - 1 & \text{if } j = i - 1 \\ 0 & \text{otherwise} \end{cases}$$

$$g_{ij} = \begin{cases} 4 & \text{if } j = 4 \\ \frac{1}{\Delta x^2} & \text{if } j = 5 \text{ and } i = 1 \\ \frac{1}{\Delta x^2} & \text{if } j = 5 \text{ and } i = N \\ 0 & \text{otherwise.} \end{cases}$$

In this case, the boundary condition and the forcing terms are given exactly, so there is no approximation error resulting from this formulation. □

Chapter 5

Bounds and Barriers for SSP Runge–Kutta Methods

In this chapter, we collect some bounds on the SSP coefficient and barriers on the order of methods with positive SSP coefficients. We are concerned with two questions:

(1) What is the highest order that an explicit or implicit Runge–Kutta method with positive SSP coefficient can attain?

(2) Can we find methods which are SSP with no time step restriction? If not, what is the optimal SSP coefficient attained by a Runge–Kutta method with a given number of stages and a given order?

Some of these results are very restrictive; for instance, we will see that implicit SSP Runge–Kutta methods have an order barrier of six, while explicit SSP Runge–Kutta methods have an order barrier of four. Furthermore, the allowable step size (or equivalently, the SSP coefficient) is relatively small. These results can be understood when we recall that the SSP property guarantees strong stability (monotonicity) in arbitrary convex functionals, for arbitrary starting values and arbitrary nonlinear, nonautonomous equations, as long as forward Euler strong stability is satisfied. This is a very strong requirement, and imposes severe restrictions on other properties of a Runge–Kutta method.

5.1 Order barriers

In Theorem 2.1, which established the SSP property for Runge–Kutta methods, we required that the coefficients α_{ij}, β_{ij} in the Shu-Osher form (2.10) be non-negative. We will show that the coefficients in the Butcher form must satisfy a similar requirement, $\mathbf{A} \geq 0$ and $\mathbf{b} > 0$, to have a non-zero SSP coefficient. Unfortunately, this requirement leads to order barriers for

SSP Runge–Kutta methods. In this section we will discuss the concept of stage order and the barriers on SSP Runge–Kutta methods. It is important to recall that these barriers do not apply to the special case of strong stability preservation for linear problems, which was discussed in Chapter 4.

5.1.1 *Stage order*

A key ingredient in deriving the order barrier for SSP Runge–Kutta methods is the concept of *stage order*.

Definition 5.1 (Stage Order). *The stage order of an m-stage method is defined as the largest integer $\tilde{p}$ for which the Butcher coefficients satisfy*

$$\sum_{j=1}^{m} b_j c_j^{k-1} = \frac{1}{k}, \qquad (1 \le k \le \tilde{p}), \qquad (5.1a)$$

$$\sum_{j=1}^{m} a_{ij} c_j^{k-1} = \frac{1}{k} c_i^k, \qquad (1 \le k \le \tilde{p}, 1 \le i \le m). \qquad (5.1b)$$

To understand the stage order, observe that each stage $u^{(i)}$ of a Runge–Kutta method is a consistent approximation to the exact solution $u(t)$ at time $t_n + c_i h$. If we assume that u^n is exact, then

$$u^{(i)} \approx u(t_n + c_i h) + h\tau_i,$$

where $\tau_i = \mathcal{O}(h^p)$ for some $p \ge 1$. By use of Taylor expansions, we can derive expressions for the local error of each stage. The result is

$$\tau_i = \sum_{k=1}^{\infty} \frac{1}{k!} \left(c_i^k - k \sum_j a_{ij} c_j^{k-1} \right) h^{k-1} u^{(k)}(t_n) . \qquad (5.2)$$

Requiring that $\tau_i = \mathcal{O}(h^p)$ leads to (5.1b). In fact, we can view u^{n+1} simply the final stage of the method; hence the same expansion holds for the final stage but with $c_i = 1$ and $a_{ij} = b_j$. This leads to (5.1a). We will refer to τ_i as the *stage truncation error*. We say stage i has order p if $\tau_i = \mathcal{O}(h^p)$; that is, if (5.1b) is satisfied for $k = 1, 2, \ldots, p - 1$. The stage order of the method is simply defined as the minimum stage order over all stages. For arbitrarily stiff problems, Runge–Kutta methods cannot in general be expected to converge at the rate indicated by their order. Instead, the rate of convergence for such problems is bounded below by the stage order.

5.1.2 Order barrier for explicit Runge–Kutta methods

Armed with the concept of *stage order*, we are ready to return to discussion of order barriers of SSP Runge–Kutta methods. We begin with a sequence of observations on the stage order and positivity of the coefficients of explicit SSP Runge–Kutta methods.

Observation 5.1. Explicit Runge–Kutta methods have stage order $\tilde{p}$ at most equal to one.

Proof. Consider the second stage $u^{(2)}$ of an explicit Runge–Kutta method. This stage can be thought of as an explicit one-stage, one step method; in other words, the second stage of an explicit Runge–Kutta method is just a forward Euler step, so this stage cannot have order greater than one. $\square$

Observation 5.2. [62] Any Runge–Kutta method with positive SSP coefficient, $\mathcal{C} > 0$, has non-negative coefficients and weights: $\mathbf{A} \geq 0, \mathbf{b} \geq 0$. If the method is irreducible, then the weights are strictly positive: $\mathbf{b} > 0$.

Proof. By Theorem 3.2, for $0 \leq r \leq \mathcal{C}$, we have that $(\mathbf{I} + r\boldsymbol{\beta}_0)$ is invertible and

$$\boldsymbol{\beta}_0(\mathbf{I} + r\boldsymbol{\beta}_0)^{-1} \geq 0 \text{ (element-wise)},$$

where

$$\boldsymbol{\beta}_0 = \begin{pmatrix} \mathbf{A} & 0 \\ \mathbf{b}^{\mathrm{T}} & 0 \end{pmatrix}.$$

Taking $r = 0$ gives $\boldsymbol{\beta}_0 \geq 0$, which implies that $\mathbf{A}, \mathbf{b} \geq 0$. Furthermore, taking $r > 0$ we have

$$\boldsymbol{\beta}_0(\mathbf{I} + r\boldsymbol{\beta}_0)^{-1} = \boldsymbol{\beta}_0 - r\boldsymbol{\beta}_0^2 + \cdots \geq 0, \tag{5.3}$$

so if $(\boldsymbol{\beta}_0)_{ij} = 0$ for some (i, j), then also $(\boldsymbol{\beta}_0^2)_{ij} = 0$. Suppose $b_j = 0$ for some j; then it must be that

$$(\boldsymbol{\beta}_0^2)_{m+1,j} = \sum_i b_i a_{ij} = 0.$$

Given the nonnegativity of $\mathbf{A}$ and $\mathbf{b}$, this means that either b_i or a_{ij} is zero for each value of i.

Now partition the set $\mathcal{S} = \{1, 2, \ldots, m\}$ into $\mathcal{S}_1, \mathcal{S}_2$ such that $b_j > 0$ for all $j \in \mathcal{S}_1$ and $b_j = 0$ for all $j \in \mathcal{S}_2$. Then $a_{ij} = 0$ for all $i \in \mathcal{S}_1$ and $j \in \mathcal{S}_2$. This implies that the method is DJ-reducible, unless $\mathcal{S}_2 = \emptyset$. $\square$

Observation 5.3. [62] For any Runge–Kutta method with $\mathbf{b} > 0$, the order p and stage order $\tilde{p}$ satisfy

$$\tilde{p} \geq \left\lfloor \frac{p-1}{2} \right\rfloor .$$

This result was proved in [62] by combining some of the order conditions. Alternatively, it can be obtained directly using Albrecht's theory [1].   □

Combining these three observations gives an order barrier on SSP Runge–Kutta methods:

Observation 5.4. Any irreducible explicit Runge–Kutta method with positive SSP coefficient has order $p \leq 4$.

The proof follows immediately from the fact that an irreducible p order Runge–Kutta method with positive SSP coefficient has $\mathbf{b} > 0$ (Observation 5.2) which implies that its stage order $\tilde{p} \geq \lfloor \frac{p-1}{2} \rfloor$ (Observation 5.3). Combining this with the fact that explicit Runge–Kutta methods have stage order $\tilde{p}$ at most equal to one (Observation 5.1), we have $1 \geq \lfloor \frac{p-1}{2} \rfloor$ and therefore $p \leq 4$.

5.1.3   *Order barrier for implicit Runge–Kutta methods*

The process of deriving an order barrier for implicit Runge–Kutta methods is very similar to that of the explicit case. However, when dealing with explicit methods, the stage order is limited whether or not one requires non-negative coefficients [62, 16, 17], whereas for implicit methods the bound on the stage order is a consequence of the requirement that $\mathbf{A} \geq 0$.

Observation 5.5. [62] Any irreducible Runge–Kutta method with non-negative coefficients $\mathbf{A} \geq 0$ must have stage order $\tilde{p} \leq 2$. Furthermore, if the method has $\tilde{p} = 2$, then $\mathbf{A}$ must have a zero row.

Proof.   First assume the method has $\mathbf{A} \geq 0$ and $\tilde{p} = 2$, and (without loss of generality) order the stages so that $0 \leq c_1 \leq c_2 \leq \cdots \leq c_m$. Then by (5.1b) we have

$$\sum_{j=1}^{m} a_{ij} = c_i \tag{5.4a}$$

$$\sum_{j=1}^{m} a_{ij} c_j = \frac{1}{2} c_i^2 . \tag{5.4b}$$

Multiplying (5.4a) by 2 and subtracting (5.4b) from the result gives

$$\frac{1}{2}c_i^2 = \sum_{j=1}^{m} a_{ij}(c_i - c_j). \tag{5.5}$$

Taking $i = 1$ in the equation above, we find that the left-hand side is non-negative, while every term of the summand on the right-hand side is non-positive. This equation can hold only if $c_1 = 0$, so the first row of $\mathbf{A}$ is zero.

Now assume $\tilde{p} > 2$. Then, in addition to (5.4), (5.1b) implies

$$\sum_{j=1}^{m} a_{ij}c_j^2 = \frac{1}{3}c_i^3. \tag{5.6}$$

Let i be the smallest index such that $c_i > 0$. Multiplying (5.4b) by c_i and subtracting (5.6) gives

$$\sum_{j=1}^{m} a_{ij}c_j(c_i - c_j) = \frac{1}{6}c_i^3. \tag{5.7}$$

Every term in the sum on the left is non-positive, while the right-hand side is positive, which is a contradiction. $\qquad\square$

Combining these observations gives the following order barrier:

Observation 5.6 (Order barrier for implicit SSP RK methods).
Any irreducible Runge–Kutta method with positive SSP coefficient must have order $p \leq 6$.

The proof follows from three observations:

(1) Irreducible Runge–Kutta methods with positive SSP coefficient, have $\mathbf{A} \geq 0$ and $\mathbf{b} > 0$ (Observation 5.1).
(2) Runge–Kutta methods with $\mathbf{A} \geq 0$ have stage order $\tilde{p}$ at most equal to two (Observation 5.5).
(3) Any Runge–Kutta method of order p with weights $\mathbf{b} > 0$ must have stage order $\tilde{p} \geq \lfloor \frac{p-1}{2} \rfloor$ (Observation 5.3). $\qquad\square$

5.1.4 Order barriers for diagonally implicit and singly implicit methods

In this subsection we consider order barriers for three special classes of implicit Runge–Kutta methods: diagonally implicit, singly implicit, and singly diagonally implicit methods.

First, consider the case where the matrix of coefficients $\mathbf{A}$ is lower triangular:

$$\mathbf{A} = \begin{bmatrix} \times & & & \\ \times & \times & & \\ \vdots & & \ddots & \\ \times & \times & \times & \times \end{bmatrix}$$

Methods in this class are referred to as diagonally implicit Runge–Kutta (DIRK) methods. Meanwhile, methods for which $\mathbf{A}$ has a single eigenvalue of multiplicity m are referred to as singly implicit Runge–Kutta (SIRK) methods. Finally, methods that are both diagonally implicit and singly implicit:

$$\mathbf{A} = \begin{bmatrix} \gamma & & & \\ \times & \gamma & & \\ \vdots & & \ddots & \\ \times & \times & \times & \gamma \end{bmatrix}$$

are known as singly diagonally implicit (SDIRK) methods.

The significance of these special cases emerges when an m-stage implicit Runge–Kutta method is applied to a system of N ODEs. This typically requires the solution of a system of mN equations. When the system results from the semi-discretization of a system of nonlinear PDEs, N is typically very large and the system of ODEs is nonlinear, making the solution of this system very expensive. Using a transformation involving the Jordan form of $\mathbf{A}$, the amount of work can be reduced [6]. This is especially efficient for *singly implicit* (SIRK) methods, because $\mathbf{A}$ has only one distinct eigenvalue so that the necessary matrix factorizations can be reused. On the other hand, *diagonally implicit* (DIRK) methods, for which $\mathbf{A}$ is lower triangular, can be implemented efficiently without transforming to the Jordan form of $\mathbf{A}$. *Singly diagonally implicit* (SDIRK) methods, which are both singly implicit and diagonally implicit (i.e. $\mathbf{A}$ is lower triangular with all diagonal entries identical), incorporate both of these advantages. The relationships between these classes are illustrated in Figure 5.1. (For details on efficient implementation of implicit Runge–Kutta methods see, e.g., [17].)

First, we state a general result on the order of DIRK and SIRK methods. This result is not directly related to their SSP properties, but it informs us what order we can expect from an m-stage method.

Observation 5.7. The order of an m-stage SIRK or DIRK method is at most $m + 1$.

Proof. For a given m-stage SIRK or DIRK method, let ψ denote the stability function. For both classes of methods, ψ is a rational function with numerator of degree m and only real poles. Such a function approximates the exponential function to order at most $m + 1$ (see Theorem 3.5.11 in [17]). □

Now we investigate order barriers for singly implicit and diagonally implicit methods with positive SSP coefficients. For singly diagonally implicit methods, the first stage is a backward Euler step (unless the method is in fact explicit), and thus at most first order. From the discussion in the sections above it is clear that SSP singly diagonally implicit Runge–Kutta methods have stage order no greater than one, and so suffer from the same order barrier ($p \le 4$) as explicit methods:

Observation 5.8 (Order barrier for SDIRK methods). *Any singly diagonally implicit Runge–Kutta method with positive SSP coefficient has order $p \le 4$ [54].*

Furthermore, the SSP coefficient of any SIRK method of order more than five cannot be very large.

Observation 5.9 (Order barrier for SIRK methods).
For m-stage SIRK methods of order $p \ge 5$, $\mathcal{C}$ is bounded by the optimal threshold factor of m-stage explicit Runge–Kutta methods of the same order [54].

5.2 Bounds on the SSP coefficient

In solving the linear autonomous initial value problem (4.1) or the nonlinear, nonautonomous initial value problem (2.2), the maximum allowable timestep that preserves strong stability is proportional to the threshold factor R or the SSP coefficient $\mathcal{C}$, respectively. Hence it is advantageous to use a method with the largest possible value of this coefficient.

We saw above that the SSP coefficient $\mathcal{C}$ of a high order ($p \ge 5$) SIRK method is bounded by the threshold factor R of an explicit method of the same number of stages and same order. In this section, we investigate similar bounds for explicit and implicit Runge–Kutta methods: we ask what is the largest possible effective SSP coefficient a Runge–Kutta method can have. For explicit methods, we will show that $\mathcal{C}_{\text{eff}} \le 1$ even in the special case of SSP methods for linear problems. For implicit methods, we ask if it is possible to obtain methods which are SSP with no time step restriction.

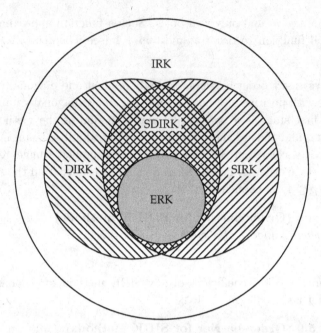

Fig. 5.1 Diagram of important classes of Runge–Kutta methods: IRK=Implicit; DIRK=Diagonally Implicit; SIRK=Singly Implicit; SDIRK=Singly Diagonally Implicit; ERK=Explicit.

It turns out that all implicit methods of order greater than one have a finite SSP coefficient. The relevant question then becomes, what is the size of the optimal SSP coefficient attained by a Runge–Kutta method with a given number of stages and a given order?

To obtain a bound on the size of the SSP coefficient, it is convenient to look at methods for linear problems and obtain an upper bound by looking at the optimal threshold factors. We focus on the effective time step as given by the *scaled threshold factor* R_{eff}. The time step restriction $\Delta t \leq \mathcal{C}$ guarantees strong stability preservation for a wider class of problems than the restriction $\Delta t \leq R$, so it follows that

$$\mathcal{C} \leq R \tag{5.8}$$

for any method. Also, given a class of methods with m stages and of order p, it follows that the optimal SSP coefficient among all methods in this class (denoted $\mathcal{C}_{m,p}$) is no larger than the optimal threshold factor: $\mathcal{C}_{m,p} \leq R_{m,p}$. For explicit Runge–Kutta methods of order $p \leq 2$, the same order conditions must be satisfied for solution of linear or nonlinear initial value problems, so that $\mathcal{C}_{m,p} = R_{m,p}$ for $p \leq 2$.

Remark 5.1. This reasoning can be generalized to linear multistep methods with s steps. In fact, it can be shown *a priori* that the SSP coefficient of an s step explicit linear multistep method is in fact equal to the threshold factor R. Furthermore, for explicit m-stage Runge–Kutta methods of order $p \leq 2$, the SSP coefficient is also equal to the threshold factor R. In addition, it turns out that $C = R$ for the *optimal* methods in some other classes. This is convenient because it is usually easier to find the optimal threshold factor than the optimal SSP coefficient over a given class of methods. If one can find the optimal threshold factor and then find a method that has SSP coefficient C equal to this value, it follows from (5.8) that this method is optimal. This reasoning can be used, for instance, to demonstrate the optimality of many SSP methods, including explicit second, third, and fourth-order Runge–Kutta methods presented in Chapters 2 and 6. $\qquad\square$

5.2.1 *Bounds for explicit Runge–Kutta methods*

We denote the optimal radius of absolute monotonicity of an explicit Runge–Kutta method with at most m stages and at least order p, $C_{m,p}^{\mathrm{ERK}}$, and recall that it is always bounded by the optimal threshold factor for m-stage, order p explicit Runge–Kutta methods. Our first observation, already stated in Chapter 4, is that the order conditions on an m-stage, order p explicit Runge–Kutta method (4.21) lead to a limit on the maximum threshold factor, $R_{m,p}^{\mathrm{ERK}} \leq m - p + 1$.

Observation 5.10. [61] The threshold factor of an explicit m-stage, pth order Runge–Kutta method is at most $m - p + 1$.

Proof. Consider the order conditions on the stability function coefficients a_{p-1} and a_p. Combining (4.21) and (4.20), these can be written as

$$\sum_{j=0}^{m} j(j-1)\cdots(j-p+2)\gamma_j = r^{p-1} \tag{5.9a}$$

$$\sum_{j=0}^{m} j(j-1)\cdots(j-p+2)(j-p+1)\gamma_j = r^{p} \tag{5.9b}$$

where all γ_j are non-negative for r less than the threshold factor of the method.

Multiply (5.9a) by $m - p + 1$ and subtract it from (5.9b) to obtain

$$\sum_{j=0}^{m} j(j-1)\cdots(j-p+2)(j-m)\gamma_j = r^{p-1}(r - m + p - 1).$$

Since

$$j(j-1)\cdots(j-p+2)(j-m) \leq 0 \text{ for } 0 \leq j \leq m,$$

then if $\gamma_j \geq 0$, the left-hand side must be negative, which implies $r \leq m - p + 1$. $\qquad\square$

For some classes of methods we have equality: $R_{m,p} = m - p + 1$, as indicated in Chapter 4. In addition, the bound $C_{m,p}^{\mathrm{ERK}} \leq R_{m,p}^{\mathrm{ERK}}$ is sharp for some classes of methods (i.e. strict equality holds), including two-stage second-order, three-stage third order, and ten-stage fourth order Runge–Kutta methods; the optimal methods in these classes were presented in Chapter 2.

Remark 5.2. In fact, we can extend the results in the previous observation to obtain an upper bound of the SSP coefficient of all methods which, when applied to a linear constant coefficient problem, result in a recurrence of the form:

$$u^{n+1} = \psi_0(\Delta t L)u^n + \psi_1(\Delta t L)u^{n-1} + \cdots + \psi_{s-1}(\Delta t L)u^{n-s+1}, \quad (5.10)$$

where each ψ_i is a polynomial of degree at most m:

$$\psi_i = \sum_{j=0}^{m} a_{ij}z^j, \qquad 1 \leq i \leq s. \tag{5.11}$$

This is an extremely wide class of methods, which includes all explicit Runge–Kutta methods of up to m stages, and all explicit linear multistep methods of up to s steps. It also include many explicit multistep multi-stage methods which have up to m stages and s steps. We will show that an m stages method (5.10) of order $p \geq 1$ has threshold factor at most equal to the number of stages: $R \leq m$ [56].

The method must satisfy the following conditions for first order accuracy:

$$\sum_{i=1}^{s} a_{i0} = 1 \tag{5.12a}$$

$$\sum_{i=1}^{s} (a_{i1} + (s - i)a_{i0}) = s. \tag{5.12b}$$

Let R denote the threshold factor of the method. Since each ψ_i is absolutely monotonic on the interval $(-R, 0]$, we can write

$$\psi_i = \sum_{j=0}^{m} \gamma_{ij} \left(1 + \frac{z}{R}\right)^j \text{ with } \gamma_{ij} \geq 0. \tag{5.13}$$

Equating the right-hand sides of Equations (5.11) and (5.13) gives the following relation between the coefficients a_{il} and γ_{ij}:

$$a_{il} = \frac{1}{l! R^l} \sum_{j=0}^{m} \gamma_{ij} \prod_{n=0}^{l-1} (j - n). \tag{5.14}$$

Substituting (5.14) in (5.12) yields

$$\sum_{i=1}^{s} \sum_{j=0}^{m} \gamma_{ij} = 1, \tag{5.15a}$$

$$\sum_{i=1}^{s} \sum_{j=0}^{m} \gamma_{ij}(j + R(s - i)) = sR. \tag{5.15b}$$

Subtracting sm times (5.15a) from (5.15b) gives

$$\sum_{i=1}^{s} \sum_{j=0}^{m} \gamma_{ij} (j + R(s - i) - sm) = s(R - m).$$

Since, for $1 \leq i \leq s$, $0 \leq j \leq m$,

$$j + R(s - i) - sm = (j - m) + R(1 - i) + (R - m)(s - 1) \leq (R - m)(s - 1),$$

we have

$$s(R - m) = \sum_{i=1}^{s} \sum_{j=0}^{m} \gamma_{ij} (j + R(s - i) - sm)$$

$$\leq (s - 1)(R - m) \sum_{i=1}^{s} \sum_{j=0}^{m} \gamma_{ij}$$

$$= (s - 1)(R - m),$$

which implies that $R \leq m$.

Since the SSP coefficient $\mathcal{C}$ is no larger than the threshold factor R, this implies that the SSP coefficient of any explicit Runge–Kutta or multistep methods, (or even any multistep multi-stage method) is at most equal to its number of stages. $\qquad\square$

5.2.2   *Unconditional strong stability preservation*

In classical numerical analysis, stability restrictions on the time step can often be avoided by use of an appropriate implicit method. For instance, A-stable Runge–Kutta methods are strongly stable in L^2 under arbitrary step size when applied to linear problems involving dissipative normal matrices. It is natural to hope that the same can be accomplished in the context of strong stability preservation in other norms. Indeed, it is easy to show that the implicit Euler method

$$u^{n+1} = u^n + \Delta t F(u^{n+1}) \qquad (5.16)$$

is unconditionally strong stability preserving [62, 35]. To see this, assume that the forward Euler condition holds:

$$\| u + \Delta t F(u) \| \leq \| u \| \text{ for } \Delta t \leq \Delta t_{\mathrm{FE}} \text{ for all } u.$$

We now rearrange the implicit Euler formula:

$$u^{n+1} = u^n + \Delta t F(u^{n+1})$$
$$(1 + \alpha)u^{n+1} = u^n + \alpha u^{n+1} + \Delta t F(u^{n+1})$$
$$u^{n+1} = \frac{1}{1+\alpha} u^n + \frac{\alpha}{1+\alpha} \left(u^{n+1} + \frac{\Delta t}{\alpha} F(u^{n+1}) \right),$$

and noting that $\| u^{n+1} + \frac{\Delta t}{\alpha} F(u^{n+1}) \| \leq \| u^{n+1} \|$ for $\frac{\Delta t}{\alpha} \leq \Delta t_{FE}$, we observe that

$$\| u^{n+1} \| \leq \frac{1}{1+\alpha} \| u^n \| + \frac{\alpha}{1+\alpha} \| u^{n+1} \|.$$

Moving all the $\| u^{n+1} \|$ terms to the left and multiplying by $(1 + \alpha)$, we conclude that $\| u^{n+1} \| \leq \| u^n \|$. Thus, this method is SSP for $\frac{\Delta t}{\alpha} \leq \Delta t_{FE}$. Note that this process is true for any value of $\alpha > 0$. Since α can take on any positive value, the implicit Euler method is unconditionally strong stability preserving.

However, for higher order methods, unconditional strong stability preservation is not possible. The result below follows from Theorem 1.3 in [95] (see also [27]).

Theorem 5.1. *If $\psi(z) - \exp(z) = \mathcal{O}(z^3)$ for $z \to 0$, then*

$$R(\psi) < \infty. \qquad (5.17)$$

Since $\mathcal{C} \leq R$, then any Runge–Kutta method of order $p > 1$ has a finite SSP coefficient:

$$p > 1 \implies \mathcal{C} < \infty. \qquad (5.18)$$

Remark 5.3. In fact, this result holds for any general linear method of the form (5.10): any such method with positive SSP coefficient will have some time step restriction. □

The question of interest becomes: what is the optimal SSP coefficient attained by a Runge–Kutta method with a given number of stages and a given order? While we do not have a theoretical answer to this question in general, numerical results (see Chapter 7) suggest that the effective SSP coefficient of any implicit Runge–Kutta method of second order or greater is at most two:

$$\mathcal{C}_{\text{eff}} \leq 2.$$

This restriction would imply that implicit SSP Runge–Kutta methods are not competitive with explicit methods in most applications, since the cost of doing an implicit solve is generally more than twice the cost of a corresponding explicit step.

Chapter 6

Low Storage Optimal Explicit SSP Runge–Kutta Methods

In Chapter 2 we presented a few optimal explicit SSP Runge–Kutta methods. The two-stage, second order, three-stage, third order, and five-stage, fourth order methods have the special property that they are the optimal schemes among those with the minimum number of stages required for a given order. Subsequent to development of these schemes, it was realized that larger SSP coefficients, and, more importantly, larger effective SSP coefficients could be achieved by methods with more than the minimum number of stages. In [97, 72, 84, 55], this strategy was pursued to achieve methods with increasingly large effective SSP coefficients. In this chapter, we review these methods.

The optimal SSP Runge–Kutta methods turn out to be optimal also in terms of the storage required for their implementation. Therefore, before presenting the optimal SSP methods themselves, we discuss in Section 6.1 low-storage Runge–Kutta methods and their relation to sparse Shu-Osher forms, as developed in [59]. In the second part of the chapter, we return to the subject of SSP methods. In Section 6.2, we review the optimal explicit SSP Runge–Kutta methods of [55], which were shown to have sparse canonical Shu-Osher forms that allow for low-storage implementation, just like the classes of methods considered in the first half of the chapter. Finally, in Section 6.3, we conclude with remarks on the embedding of optimal SSP pairs.

For further details regarding the methods in this chapter, including analysis of their linear stability properties and error coefficients, see [55]. For more details on the low-storage algorithms, see [59].

6.1   Low-storage Runge–Kutta algorithms

The search for SSP explicit Runge–Kutta methods with large allowable time step focuses on efficiency in the sense of minimizing the computational cost involved in evolving a system of ODEs to some final time. However, there is another type of efficiency which may need to be considered: the computational memory required for each step of the time evolution. A naive implementation of an m-stage Runge–Kutta method requires $m + 1$ memory registers.

In early computers, only a very small amount of memory was available for storing both the program and intermediate results of the computation. This constraint led Gill to devise a four-stage fourth order Runge–Kutta method that could be implemented using only three memory registers [23]. The method relied on a particular algebraic relation between the coefficients, such that only certain combinations of previous stages (rather than all of the stages themselves) needed to be stored. This is the basic idea underlying all low-storage Runge–Kutta methods.

On modern computers, storage space for programs is no longer such a pressing concern; however, when integrating very large numbers of ODEs, fast memory for temporary storage during a computation is often the limiting factor. This is typically the case in method-of-lines discretizations of PDEs, and modern efforts have focused on finding low-storage methods that are also optimized for stability and/or accuracy relative to particular semi-discretizations of PDEs.

Here we look for ways of implementing high order Runge–Kutta methods that are SSP and that have low-storage properties. It is fortunate that optimal SSP methods turn out to have very efficient low-storage implementations. In fact, we will see in the next section that these efficient implementations correspond to the canonical Shu-Osher form introduced in Section 3.3, with $r = \mathcal{C}$.

Let N denote the size of the system of ODEs to be integrated. In systems resulting from a method-of-lines discretization of PDEs, this is typically the number of PDEs multiplied by the number of gridpoints. Then we say a Runge–Kutta method requires M registers if each step can be calculated using $MN + o(N)$ memory locations.

Let S_1, S_2 represent two N-word registers in memory. Then it is always assumed that we can make the assignments

$$S_1 := F(S_2)$$

and

$$S_1 := c_1 S_1 + c_2 S_2$$

without using additional memory beyond these two registers. Here $a := b$ means 'the value of b is stored in a'. Using only these two types of assignments, it is straightforward to implement an m-stage method using $m + 1$ registers.

Three low-storage Runge–Kutta algorithms have been proposed that require only two registers [110, 52, 55]. Each requires some additional type of assignment, and each is related to a certain sparse Shu-Osher form. In this section, we discuss these three algorithms, and the particular assumptions they make regarding the evaluation and storage of F.

6.1.1 *Williamson (2N) methods*

Williamson methods [110] can be implemented as follows:

Algorithm 6.1 (Williamson (2N)).

$$(y_1) \quad S_1 := u^n$$
$$\text{for } i = 2 : m + 1 \text{ do}$$
$$S_2 := A_i S_2 + \Delta t F(S_1)$$
$$(y_i) \quad S_1 := S_1 + B_i S_2$$
$$\text{end}$$
$$u^{n+1} = S_1$$

with $A_2 = 0$.

The coefficients above are related to the Butcher coefficients as follows:

$$B_i = a_{i+1,i} \qquad\qquad i < m$$
$$B_m = b_m$$
$$A_i = \frac{b_{i-1} - a_{i,i-1}}{b_i} \qquad\qquad b_i \neq 0$$
$$A_i = \frac{a_{i+1,i-1} - c_i}{a_{i+1,i}} \qquad\qquad b_i = 0.$$

Williamson methods require only 2 memory registers, if we make the following assumption on the computation of F:

Assumption 6.1 (Williamson). *Assignments   of   the   form* $S_2 := S_2 + F(S_1)$ *can be made with only* $2N + o(N)$ *memory.*

Note that an m-stage method of this form has $2m - 1$ free parameters. By straightforward algebraic manipulation, Williamson methods can be written in a Shu-Osher form (2.10) where all the internal stages have the form

$$u^{(i)} = \alpha_{i,i-2} u^{(i-2)} + (1 - \alpha_{i,i-2}) u^{(i-1)} + \beta_{i,i-1} \Delta t F(u^{(i-1)}). \qquad (6.1)$$

Here and elsewhere, any coefficients with negative indices are taken to be zero, so that α is bidiagonal and β is diagonal. Thus it turns out that Williamson methods possess a Shu-Osher form in which the matrices α, β are very sparse.

6.1.2   van der Houwen (2R) methods

A different low-storage algorithm was developed by van der Houwen [105] and Wray [52].

Algorithm 6.2 (van der Houwen (2R)).

$$S_2 := u^n$$

$\quad\quad$ for i = 1 : m do

(y_i) $\quad\quad$ $S_1 := S_2 + (a_{i,i-1} - b_{i-1})\Delta t S_1$

$\quad\quad\quad\quad$ $S_1 := F(S_1)$

$\quad\quad\quad\quad$ $S_2 := S_2 + b_i \Delta t S_1$

$\quad\quad$ end

$\quad\quad\quad$ $u^{n+1} = S_2.$

The coefficients a_{ij}, b_j *are the Butcher coefficients, and we define* $a_{10} = b_0 = 0$.

As in the case of Williamson methods, an m-stage method has $2m - 1$ free parameters. In order to implement these methods with just two registers, the following assumption is required:

Assumption 6.2 (van der Houwen). *Assignments   of   the   form* $S_1 := F(S_1)$ *can be made with only* $N + o(N)$ *memory.*

In [52], an implementation is given that requires swapping the roles of the two registers at each stage. Here we have followed the implementation of [10] as it is less complicated in that the roles of the two registers need not be swapped at each stage. Methods of van der Houwen type have also been proposed in [103]. The low-storage methods of [45] can be viewed as a subclass of van der Houwen methods with especially simple structure.

Once again, the Shu-Osher form of these methods is sparse. Observe that van der Houwen methods can be written in a Shu-Osher form with the internal stages given by

$$u^{(i)} = u^{(i-1)} + \beta_{i,i-2}\Delta t F(u^{(i-2)}) + \beta_{i,i-1}\Delta t F(u^{(i-1)}), \qquad (6.2)$$

so that α is diagonal and β is bidiagonal.

6.1.3 *2S and 2S* methods*

Based on the Shu-Osher forms presented above for 2N and 2R methods, it is natural to ask whether it is possible to implement a method with just two registers if α and β have other types of sparse structure. Perhaps the most obvious generalization is to allow both matrices to be bidiagonal, i.e.

$$u^{(i)} = \alpha_{i,i-2}u^{(i-2)} + (1 - \alpha_{i,i-2})u^{(i-1)} + \beta_{i,i-2}\Delta t F(u^{(i-2)}) \qquad (6.3)$$
$$+\beta_{i,i-1}\Delta t F(u^{(i-1)}).$$

It turns out that this is possible, under the following assumption, which was introduced in [55]:

Assumption 6.3 (Ketcheson). *Assignments of the form*

$$S_1 := S_1 + F(S_1) \qquad (6.4)$$

can be made with only $N + o(N)$ memory.

Examining the Shu-Osher form (6.3), it is clear that 2S methods may be implemented (under Assumption 6.4) using two registers if one is willing to evaluate $F(u^{(i)})$ twice for each stage i. With a little care, however, this doubling of the number of function evaluations can be avoided. The resulting algorithm is:

Algorithm 6.3 (Ketcheson (2S)).

$$S_2 := 0$$

(y_1) $S_1 := u^n$

for $i = 2 : m + 1$ do

$$S_2 := S_2 + \delta_{i-1}S_1$$

(y_i) $S_1 := \gamma_{i1}S_1 + \gamma_{i2}S_2 + \beta_{i,i-1}\Delta t F(S_1)$

end

$$u^{n+1} = S_1$$

The value of δ_m makes no essential difference (any change can be compensated by changing $\gamma_{m+1,1}, \gamma_{m+1,2}$), so we set it to zero. Consistency requires that $\delta_1 = 1, \gamma_{22} = 1$, and

$$\gamma_{i,1} = 1 - \gamma_{i,2}\sum_{j=1}^{i-1}\delta_j \qquad 2 \leq i \leq m + 1,$$

leaving $3s - 3$ free parameters – significantly more than for the 2N or 2R methods. We refer to these methods as 2S methods. In the next section we present explicit SSP Runge–Kutta methods of this type which are of order $p = 2, 3, 4$.

It is common to check some accuracy or stability condition after each step, and to reject the step if the condition is violated. In this case, the solution from the last timestep, u^n, must be retained during the computation of u^{n+1}. For 2R/2N/2S methods, this will require an additional register. On the other hand, in [55, 59], methods were proposed that can be implemented using only two registers, with one register retaining the previous solution. We refer to these as 2S* methods. These methods have Shu-Osher form

$$y_i = \alpha_{i,1}u^n + \alpha_{i,i-1}y_{i-1} + \beta_{i,i-1}\Delta t F(y_{i-1})$$

the low-storage implementation of such methods (Algorithm 6.4), is straightforward, given the sparse Shu-Osher form. It is equivalent to the 2S algorithm above, but with $\gamma_{i2} = \alpha_{i,1}$ and $\delta_i = 0$ except $\delta_1 = 1$.

Algorithm 6.4 (Ketcheson (2S*)).

(y_1) $S_1 := u^n$ $S_2 := u^n$

   for $i = 2 : m + 1$ do

(y_i) $S_1 := (1 - \alpha_{i,1})S_1 + \alpha_{i,1}S_2 + \beta_{i,i-1}\Delta t F(S_1)$

   end

   $u^{n+1} = S_1$

6.2   Optimal SSP low-storage explicit Runge–Kutta methods

In this section, we review explicit Runge–Kutta methods that are optimal in the sense of having the largest SSP coefficient among all methods for a given order and number of stages. We will see that the optimal SSP methods, like the low-storage methods considered in the previous section, have sparse Shu-Osher forms. Using the assumption (6.4), they can be implemented using very little storage. This material appeared originally in [55].

We showed in Chapter 5 that any explicit Runge–Kutta method must have effective SSP coefficient $\mathcal{C}_{\text{eff}} \leq 1$. Furthermore, any method that uses only a single memory register must consist simply of repeated forward Euler steps and therefore cannot be more than first order accurate. Thus an ideal explicit low-storage SSP method of higher order would use two storage registers and have $\mathcal{C}_{\text{eff}}$ as close as possible to unity. Since explicit SSP Runge–Kutta methods have order at most four (as we saw in Chapter 5), we consider methods of order two through four. Remarkably, we will see that this ideal can be achieved for second and third order methods.

In Table 6.1 we present a comparison between some of the optimal methods presented in Chapter 2 and those to be presented in this chapter.

6.2.1   *Second order methods*

Optimal second order methods with $\mathcal{C}_{\text{eff}}$ arbitrarily close to 1 were found in [62], and later independently in [97]. We have discussed these methods already in Chapter 4. Here we present their low-storage 2S* implementation.

Table 6.1 Properties of popular and of optimal explicit SSP Runge–Kutta methods. An asterisk indicates that the previous time step can be retained without increasing the required number of registers.

Popular Method	$\mathcal{C}_{\text{eff}}$	Storage	Improved Method	$\mathcal{C}_{\text{eff}}$	Storage
SSPRK(2,2)	0.500	2S*	SSPRK(m,2)	$1 - 1/m$	2S*
SSPRK(3,3)	0.333	2S*	SSPRK(4,3)	0.500	2S*
			SSPRK(n^2,3)	$1 - 1/n$	2S
SSPRK(5,4)	0.377	3S	SSPRK(10,4)	0.600	2S

Low-storage SSPRK(m,2): The m-stage method in this family has SSP coefficient $m - 1$, hence $\mathcal{C}_{\text{eff}} = \frac{m-1}{m}$. The non-zero entries of the low-storage form for the m-stage method are

$$\beta_{i,i-1} = \begin{cases} \frac{1}{m-1} & 1 \le i \le m-1 \\ \frac{1}{m} & i = m \end{cases}$$

$$\alpha_{i,i-1} = \begin{cases} 1 & 1 \le i \le m-1 \\ \frac{m-1}{m} & i = m \end{cases}$$

$$\alpha_{m,0} = \frac{1}{m}.$$

The abscissas are

$$c_i = \frac{i - 1}{m - 1} \quad (1 \le i \le m). \tag{6.5}$$

A low-storage implementation is given in program 6.1. The storage registers in the program are denoted by q_1 and q_2.

```
Program 6.1 (Low-storage second order RK).
q1 = u;  q2 = u;
for    i=1:m-1
       q1 = q1 + dt*F(q1)/(m-1);
end
q1 = ( (m-1)*q1 + q2 + dt*F(q1) )/m;
u=q1;
```

Note that, as 2S* methods, these methods do not require a third register even if the previous time step must be retained. Because the storage costs do not increase with m while the effective SSP coefficient does, there seems to be little drawback to using these methods with large values of m. For large values of m, members of this family are approximately twice as efficient as the two-stage method, which is the most commonly used.

6.2.2 Third order methods

The three- and four-stage third order SSP Runge–Kutta methods, originally reported in [92] and [62], admit a 2S* implementation [55]. In each case, it is necessary to store only the previous solution value u^n and the most recently computed stage.

The three-stage third order method has $\mathcal{C} = 1$ and $\mathcal{C}_{\text{eff}} = \frac{1}{3}$.

```
Program 6.2 (Low-storage SSPRK(3,3)).
q1 = u;
q2 = q1 + dt*F(q1);
q2 = 3/4 * q1 + 1/4 * (q2+F(q2))
q2 = 1/3 * q1 + 2/3 * (q2+F(q2))
u = q2
```

The three-stage fourth order method has $\mathcal{C} = 2$ and $\mathcal{C}_{\text{eff}} = \frac{1}{2}$.

```
Program 6.3 (Low-storage SSPRK(4,3)).
q1 = u;
q2 = q1 + dt/2 * F(q1);
q2 = q2 + dt/2 * F(q2)
q2 = 2/3 * q1 + 1/3 * (q2 + dt/2 * F(q2))
q2 = q2 + dt/2 * F(q2)
u = q2
```

Since the four-stage method is 50% more efficient and requires the same amount of memory, it seems always preferable to the three-stage method.

Larger effective SSP coefficients can again be attained by using additional stages. A family of methods with $\mathcal{C}_{\text{eff}}$ approaching the ideal value of 1 was introduced in [55].

Low-storage SSPRK(n^2,3): Let $n > 2$ be an integer and let $m = n^2$. The m-stage method given by

$$\alpha_{i,i-1} = \begin{cases} \frac{n-1}{2n-1} & \text{if } i = \frac{n(n+1)}{2} \\ 1 & \text{otherwise} \end{cases}$$

$$\alpha_{\frac{n(n+1)}{2}, \frac{(n-1)(n-2)}{2}} = \frac{n}{2n-1}$$

$$\beta_{i,i-1} = \frac{\alpha_{i,i-1}}{n^2 - n},$$

has an SSP coefficient $\mathcal{C} = R_{m,1,3} = n^2 - n$, and an effective $\mathcal{C}_{\text{eff}} = 1 - \frac{1}{n}$. The abscissas of the method are

$$c_i = \frac{i-1}{n^2-n}, \qquad \text{for } 1 \le i \le n(n+1)/2$$
$$c_i = \frac{i-n-1}{n^2-n} \qquad \text{for } n(n+1)/2 < i \le n^2.$$

These methods are optimal in the sense that no third order s-stage method exists with a larger SSP coefficient. The optimality follows from (5.8) and the fact that $R_{n^2,1,3} = n^2 - n$ [60].

Like the second order family (6.5) above, this family of methods achieves effective SSP coefficients arbitrarily close to one while using only two memory registers. Also, like the family (6.5), and unlike most known optimal third order SSP methods, the coefficients are simple rational numbers. Note that the four-stage 2S* method discussed above is the first member of this family. However, the other methods in this family cannot be implemented in 2S* form. A 2S implementation is given below.

Program 6.4 (Low-storage third order RK).

```
n = sqrt(m); r = m-n; q1 = u;
for i=1:(n-1)*(n-2)/2
   q1 = q1 + dt*F(q1)/r;
end
q2=q1;
for i=(n-1)*(n-2)/2+1:n*(n+1)/2-1
   q1 = q1 + dt*F(q1)/r;
end
q1 = ( n*q2 + (n-1)*(q1 + dt*F(q1)/r) ) / (2*n-1);
for i=n*(n+1)/2+1:m
   q1 = q1 + dt*F(q1)/r;
end
u=q1;
```

As an example, for $n = 3, m = 9$, we have the method

$$\beta_{i,i-1} = \begin{cases} \frac{1}{6} & i \in \{1..5, 7..9\} \\ \frac{1}{15} & i = 6 \end{cases}, \quad \alpha_{i,i-1} = \begin{cases} 1 & (i \in \{1..5, 7..9\} \\ \frac{2}{5} & i = 6 \end{cases}, \quad \alpha_{6,1} = \frac{3}{5}.$$

6.2.3 *Fourth order methods*

As seen in Chapter 5, no explicit fourth order method with four stages has $\mathcal{C} > 0$. The five-stage fourth order method given in Chapter 2 requires 3 registers. A 2S fourth order SSP method was given in [55]. It is a ten-stage fourth order method implementable with two registers (see Algorithm 6.5) and with an effective SSP coefficient greater than any previously known fourth order method. Additionally, it is the only fourth order SSP method to be analytically proved optimal, because it achieves the optimal bound on ten-stage, fourth order SSP methods for linear problems: $\mathcal{C} = R_{10,4} = 6$. Finally, the method has simple rational coefficients. The non-zero coefficients are

$$\beta_{i,i-1} = \begin{cases} \frac{1}{6} & i \in \{1..4, 6..9\} \\ \frac{1}{15} & i = 5 \\ \frac{1}{10} & i = 10 \end{cases}, \quad \alpha_{i,i-1} = \begin{cases} 1 & i \in \{1..4, 6..9\} \\ \frac{2}{5} & i = 5 \\ \frac{3}{5} & i = 10 \end{cases}$$

$$\beta_{10,4} = \frac{3}{50}, \quad \alpha_{5,0} = \frac{3}{5}, \quad \alpha_{10,0} = \frac{1}{25}, \quad \alpha_{10,4} = \frac{9}{25}.$$

The abscissas are

$$c = \frac{1}{6} \cdot (0, 1, 2, 3, 4, 2, 3, 4, 5, 6)^T. \tag{6.6}$$

Program 6.5 (Low-storage SSPRK(10,4)).

```
q1 = u; q2=u;
for i=1:4
  q1 = q1 + dt*F(q1)/6;
end
q2 = 1/25*q2 + 9/25*q1;
q1 = 15*q2-5*q1;
for i=6:9
  q1 = q1 + dt*F(q1)/6;
end
q1 = q2 + 3/5*q1 + 1/10*dt*F(q1);
u=q1;
```

6.3   Embedded optimal SSP pairs

Embedded Runge–Kutta pairs provide an estimate of the local truncation error that can be computed at little cost. This fact makes them useful for automatic error control. When using an SSP method to enforce an important constraint, it is very desirable to have an embedded method that is also SSP under the same (or larger) time step restriction, since violation of the constraint in the computation of the error estimate might lead to adverse effects. Macdonald [72] showed that it is possible to embed pairs of SSP methods. It turns out that it is also possible to create embedded pairs from some of the optimal low-storage SSP methods. Below we give two examples of embedded pairs.

Example 6.1. The optimal three-stage second order method and optmal four-stage third order method can be combined as an embedded pair as follows:

$$u^{(0)} = u^n$$

$$u^{(1)} = u^{(0)} + \frac{\Delta t}{2} F(u^{(0)})$$

$$u^{(2)} = u^{(1)} + \frac{\Delta t}{2} F(u^{(1)})$$

$$u_2^{n+1} = \frac{1}{3} u^{(0)} + \frac{2}{3} \left(u^{(2)} + \frac{\Delta t}{2} F(u^{(2)}) \right)$$

$$u^{(3)} = \frac{2}{3} u^{(0)} + \frac{1}{3} \left(u^{(2)} + \frac{\Delta t}{2} F(u^{(2)}) \right)$$

$$u_3^{n+1} = u^{(3)} + \frac{\Delta t}{2} F(u^{(3)}) .$$

Here u_2^{n+1}, u_3^{n+1} are the second and third order approximations corresponding to SSPRK(3,2) and SSPRK(4,3). Note that no extra function evaluations are required – the lower order method is free. □

Example 6.2. The seven-stage second order method (SSPRK(7,2)) can be used as an embedded method with the nine-stage third order method

(SSPRK(9,3)) for error control as follows:

$$u^{(0)} = u^n$$

$$u^{(i)} = u^{(i-1)} + \frac{\Delta t}{6} F(u^{(i-1)}) \quad 1 \le i \le 6$$

$$u_2^{n+1} = \frac{1}{7} \left(u^n + 6u^{(6)} + \frac{\Delta t}{6} F(u^{(6)}) \right)$$

$$u^{(6)*} = \frac{3}{5} u^{(1)} + \frac{2}{5} u^{(6)}$$

$$u^{(i)*} = u^{(i-1)*} + \frac{\Delta t}{6} F(u^{(i-1)*}) \quad 7 \le i \le 9$$

$$u_3^{n+1} = u^{(9)*}.$$

Here u_2^{n+1}, u_3^{n+1} are the second and third order approximations corresponding to SSPRK(7,2) and SSPRK(9,3), respectively. Note that, in this case, one extra function evaluation is required over what is necessary for SSPRK(9,3) alone. □

Chapter 7

Optimal Implicit SSP Runge–Kutta Methods

The optimal Runge–Kutta methods presented in Chapters 2, 4, and 6 are all of explicit type. The restrictive bounds on the SSP coefficient and order of explicit SSP Runge–Kutta methods suggests consideration of SSP implicit Runge–Kutta methods, in order to find methods with larger SSP coefficients or methods of order higher than four. In this chapter, we review results in this direction that appeared originally in [54].

In the case of classical stability properties (such as linear stability or B-stability), implicit methods exist that are stable under arbitrarily large time steps, so it is natural to hope that the same can be accomplished in the context of strong stability preservation. Unfortunately, as discussed already in Chapter 5, this proves impossible for all but first order methods.

It can be easily shown that any spatial discretization F which is strongly stable in some norm using the forward Euler method with some finite time step restriction will be *unconditionally* strongly stable, in the same norm, using the implicit Euler method [62, 49, 35]. To see this, we begin with the forward Euler method

$$u^{n+1} = u^n + \Delta t F(u^n)$$

and assume that it is strongly stable $\|u^{n+1}\| \leq \|u^n\|$ under the time step restriction $\Delta t \leq \Delta t_{FE}$. We now rearrange the implicit Euler formula

$$u^{n+1} = u^n + \Delta t F(u^{n+1})$$

$$(1 + \alpha)u^{n+1} = u^n + \alpha u^{n+1} + \Delta t F(u^{n+1})$$

$$u^{n+1} = \frac{1}{1 + \alpha}u^n + \frac{\alpha}{1 + \alpha}\left(u^{n+1} + \frac{\Delta t}{\alpha}F(u^{n+1})\right),$$

for any $\alpha > 0$. We can now bound the norm (or semi-norm, or convex functional) of the solution

$$\|u^{n+1}\| \leq \frac{1}{1 + \alpha}\|u^n\| + \frac{\alpha}{1 + \alpha}\|u^{n+1} + \frac{\Delta t}{\alpha}F(u^{n+1})\|.$$

Observing that

$$\|u^{n+1} + \frac{\Delta t}{\alpha} F(u^{n+1})\| \leq \|u^{n+1}\|$$

for $\frac{\Delta t}{\alpha} \leq \Delta t_{FE}$ we get

$$\|u^{n+1}\| \leq \frac{1}{1+\alpha} \|u^n\| + \frac{\alpha}{1+\alpha} \|u^{n+1}\|$$

so that

$$\left(1 - \frac{\alpha}{1+\alpha}\right) \|u^{n+1}\| \leq \frac{1}{1+\alpha} \|u^n\|,$$

and therefore $\|u^{n+1}\| \leq \|u^n\|$. Thus, the implicit Euler method is SSP for $\frac{\Delta t}{\alpha} \leq \Delta t_{FE}$. Note that this process is true for any value of $\alpha > 0$, so the implicit Euler method is unconditionally strong stability preserving. However, as we saw in Chapter 5, unconditional strong stability preservation is not possible for methods of greater than first order.

The following example illustrates the disadvantage of using a non-SSP method, even if the method has other good stability properties.

Example 7.1. Consider Burgers' equation $u_t + \left(\frac{1}{2}u^2\right)_x = 0, \quad x \in [0,2)$ with initial condition $u(0,x) = \frac{1}{2} - \frac{1}{4}\sin(\pi x)$ and periodic boundary conditions. We discretize in space using the conservative upwind approximation

$$-f(u)_x \approx F(u) = -\frac{1}{\Delta x} \left(f(u_i) - f(u_{i-1})\right)$$

where $f(u_i) = \frac{1}{2}u_i^2$. This spatial discretization is total variation diminishing (TVD) for $\Delta t \leq \Delta x$ when coupled with forward Euler, and unconditionally TVD (as shown above) when coupled with backward Euler.

For the time discretization, we use the second order implicit trapezoidal rule

$$u^{n+1} = u^n + \frac{1}{2}\Delta t \left(F(u^{n+1}) + F(u^n)\right)$$

which is A-stable, like the backward Euler method. Hence the full discretization has nice linear stability properties: it is absolutely stable under any time step. Nevertheless, we find that for $\Delta t > 2\Delta x$, oscillations appear in the numerical solution, as seen in Figure 7.1 on the left. This is not unexpected, since the apparent SSP coefficient for this method is $\mathcal{C} = 2$. The combination of the spatial discretization with the implicit trapezoidal rule is only TVD when the SSP time step restriction is satisfied.

A similar behavior can be observed when we use the implicit midpoint rule

$$u^{(1)} = u^n + \frac{\Delta t}{2} F(u^{(1)}) \tag{7.1}$$

$$u^{n+1} = u^{(1)} + \frac{\Delta t}{2} F(u^{(1)})$$

which is A-stable, L-stable, and B-stable, but (as seen in Figure 7.1, right) not SSP for $\Delta t > 2\Delta x$.

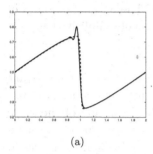

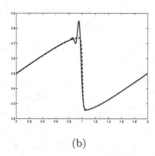

(a)   (b)

Fig. 7.1   Numerical solution of Burgers' equation with forward difference in space and the implicit trapezoid rule (left) or implicit midpoint rule (right).

□

As proved already in Chapter 5, no Runge–Kutta method of order greater than one can be unconditionally SSP, even if only linear problems are considered. To obtain methods that are unconditionally SSP, we must go outside the class of general linear methods of the type defined in Remark 5.2. This can be accomplished, for example, by using diagonally split Runge–Kutta methods [73]. However, an implicit SSP Runge–Kutta method may be useful if it has a large enough SSP coefficient. Hence an important question is whether the allowable SSP step size can be large enough to offset the extra computational effort required in the implicit solution of the resulting system at each iteration. In the following sections we will review the order barriers and SSP bounds of implicit Runge–Kutta methods presented in Chapter 5, discuss other advantages of implicit SSP Runge–Kutta methods, and present some of the optimal SSP methods found.

7.1   Barriers, bounds, and bonuses

As mentioned in Chapter 5, implicit SSP methods cannot have order greater than six. Methods of order up to five were constructed in [62]; existence of sixth order methods was established in [54], demonstrating that the order barrier is sharp. In Chapter 5, we also mentioned the fact that singly diagonally implicit SSP methods have order at most four. SSP methods in this class were investigated in [22], including optimal methods (found by numerical optimization) of up to order four and up to eight stages. In [54], fully implicit SSP Runge–Kutta methods were investigated. Remarkably, searching among the class of fully implicit methods, the optimal methods of second and third order were found to be singly diagonally implicit; in fact, they were the very methods found already in [22]. The optimal methods of fourth through sixth order were found to be diagonally implicit. Unfortunately, all of these methods turn out to have effective SSP coefficients less than or equal to two.

The SSP time step restriction provides a guarantee of other important properties. When considering implicit Runge–Kutta methods, it is important to determine whether there exists a unique solution of the stage equations. The strong stability preserving time step restriction turns out to be sufficient for this as well [62]. Furthermore, the SSP condition serves to guarantee that the errors introduced in the solution of the stage equations due to numerical roundoff and (for implicit methods) errors in the implicit solver are not unduly amplified [62].

In the following sections, we present the optimal methods in the modified Shu-Osher form (3.7)

$$u^{(i)} = v_i u^n + \sum_{j=1}^{m} \left(\alpha_{ij} u^{(j)} + \Delta t \beta_{ij} F(u^{(j)}) \right) \quad (1 \le i \le m + 1)$$

$$u^{n+1} = u^{(m+1)}.$$

To simplify implementation, we use the canonical Shu-Osher form of Chapter 3 with $r = \mathcal{C}$, and then rearrange so that the diagonal elements of $\boldsymbol{\alpha}$ are zero. For DIRK methods, as with explicit methods, we define the *effective SSP coefficient* of an m-stage method by $\mathcal{C}_{\text{eff}} = \frac{\mathcal{C}}{m}$; this normalization enables us to compare the cost of integration up to a given time using DIRK schemes of order $p > 1$. However, for non-DIRK methods with different numbers of stages, it is much less obvious how to compare computational cost. Fortunately, all the optimal methods found are diagonally implicit. The coefficients of the optimal methods are presented in Section 7.4.

7.2 Optimal second order and third order methods

Remarkably, the optimal second order and third order implicit SSP Runge–Kutta methods are singly diagonally implicit. These methods were found in [20, 22] in a numerical search for optimal singly diagonally implicit Runge–Kutta methods, and again in [54] in a search among *all* implicit Runge–Kutta methods.

Several of these methods are known to be globally optimal. The one-stage and two-stage second order methods were analytically proven optimal in [22], and the three-stage second order method was computationally proven optimal using BARON in [54]. The two-stage third order method achieves the optimal threshold factor $R(\psi)$ found in [104] for ψ in the set of third order rational approximations to the exponential with numerator and denominator of degree at most two. Since the corresponding one-parameter optimization problem is easy to solve, then (since $\mathcal{C} \leq R$), the method is clearly optimal to within numerical precision. BARON provided a certificate of global optimality [54] for the three-stage third order method (7.3). The other methods do not have a guarantee of global optimality; however because numerical searches recovered all known globally optimal methods, it seems likely that other methods found by such search – especially those of lower order or fewer stages – may be globally optimal.

Implicit SSPRK (m,2): The numerically optimal second order method with m stages is

$$
\alpha = \begin{bmatrix} 0 & & & \\ 1 & 0 & & \\ & 1 & \ddots & \\ & & \ddots & 0 \\ & & & 1 \end{bmatrix}, \quad
\beta = \begin{bmatrix} \frac{1}{2m} & & & \\ \frac{1}{2m} & \frac{1}{2m} & & \\ & \frac{1}{2m} & \ddots & \\ & & \ddots & \frac{1}{2m} \\ & & & \frac{1}{2m} \end{bmatrix}. \tag{7.2}
$$

These methods have SSP coefficient $\mathcal{C} = 2m$, and effective SSP coefficient $\mathcal{C}_{\text{eff}} = 2$. Note the sparse, bidiagonal modified Shu-Osher arrays, which make these methods efficient to implement. The one-stage method of this class is the implicit midpoint rule, while the m-stage method is equivalent to m successive applications of the implicit midpoint rule [20]. Thus these methods inherit the desirable properties of the implicit midpoint rule such as algebraic stability and A-stability [32]. If these methods are indeed optimal, this would imply that the effective SSP coefficient of any Runge–Kutta method of order greater than one is at most equal to two.

Table 7.1 Effective SSP coefficients of best known explicit and implicit methods. A dash indicates that SSP methods of this type cannot exist, a blank space indicates none were found.

	Explicit Methods			Implicit Methods				
$m \setminus p$	2	3	4	2	3	4	5	6
1	-	-	-	2	-	-	-	-
2	0.5	-	-	2	1.37	-	-	-
3	0.67	0.33	-	2	1.61	0.68	-	-
4	0.75	0.5	-	2	1.72	1.11	0.29	
5	0.8	0.53	0.30	2	1.78	1.21	0.64	
6	0.83	0.59	0.38	2	1.82	1.30	0.83	0.030
7	0.86	0.61	0.47	2	1.85	1.31	0.89	0.038
8	0.88	0.64	0.52	2	1.87	1.33	0.94	0.28
9	0.89	0.67	0.54	2	1.89	1.34	0.99	0.63
10	0.9	0.68	0.60	2	1.90	1.36	1.01	0.81
11	0.91	0.69	0.59	2	1.91	1.38	1.03	0.80

Implicit SSPRK (m,3):   The numerically optimal third order implicit Runge–Kutta methods with two or more stages are given by:

$$
\beta = \begin{bmatrix} \beta_{11} & & & \\ \beta_{21} & \ddots & & \\ & \ddots & \beta_{11} & \\ & & \beta_{21} & \beta_{11} \\ & & & \beta_{m+1,m} \end{bmatrix}, \quad
\alpha = \begin{bmatrix} 0 & & & \\ 1 & \ddots & & \\ & \ddots & 0 & \\ & & 1 & 0 \\ & & & \alpha_{m+1,m} \end{bmatrix}, \quad (7.3)
$$

where

$$
\beta_{11} = \frac{1}{2}\left(1 - \sqrt{\frac{m-1}{m+1}}\right), \quad \beta_{m+1,m} = \frac{m+1}{m(m+1+\sqrt{m^2-1})},
$$

$$
\beta_{21} = \frac{1}{2}\left(\sqrt{\frac{m+1}{m-1}} - 1\right), \quad \alpha_{m+1,m} = \frac{(m+1)(m-1+\sqrt{m^2-1})}{m(m+1+\sqrt{m^2-1})}.
$$

These methods have SSP coefficient $\mathcal{C} = m - 1 + \sqrt{m^2 - 1}$. The $m = 2$ method was shown to be optimal in [22].

7.3   Optimal fourth through sixth order methods

Based on these results, one might expect all optimal implicit SSP methods to be singly diagonally implicit. This is not possible, because according to Chapter 5, Observation 5.9, SSP SDIRK methods cannot have order greater than 4. The numerically optimal methods of fourth order are not singly diagonally implicit either; however, all numerically optimal, order four through six methods found are diagonally implicit, despite the fact that the search was conducted over all fully implicit methods.

We list the effective SSP coefficients of the numerically optimal methods in Table 7.1, and also list the optimal effective SSP coefficients for explicit methods for comparison. In this table, a dash indicates that SSP methods of this type cannot exist, and a blank space indicates that no SSP methods of this type were found. We note that $C_{\text{eff}} \leq 2$ for all methods. Compared to the higher order numerically optimal SDIRK found in [22], our numerically optimal DIRK methods have larger SSP coefficients in every case. Furthermore, they have representations that allow for very efficient implementation in terms of storage. However, SDIRK methods may be implemented in a potentially more efficient manner (in terms of computation) than DIRK methods, so the relative efficiencies of these methods will depend on their implementation. The coefficients of these methods are given in the next section.

7.4 Coefficients of optimal implicit SSP Runge–Kutta methods

7.4.1 *Fourth order methods*

7.4.1.1 *Coefficients of the optimal 3-stage 4th order implicit SSP RK*

$\alpha_{21} = 0.703541497995214$

$\alpha_{32} = 0.694594303739345$ $\beta_{11} = 0.157330905682085$ $\beta_{33} = 0.157021682372699$

$\alpha_{41} = 0.168078141811591$ $\beta_{21} = 0.342491639470766$ $\beta_{41} = 0.081822264233578$

$\alpha_{42} = 0.162500172803529$ $\beta_{22} = 0.047573123554705$ $\beta_{42} = 0.079106848361263$

$\alpha_{43} = 0.549902549377947$ $\beta_{32} = 0.338136048168635$ $\beta_{43} = 0.267698531248384$

7.4.1.2 *Coefficients of the optimal 4-stage 4th order implicit SSP RK*

$\alpha_{21} = 1$

$\alpha_{32} = 0.799340893504885$ $\beta_{11} = 0.119309657880174$ $\beta_{43} = 0.212545672537219$

$\alpha_{43} = 0.939878564212065$ $\beta_{21} = 0.226141632153728$ $\beta_{44} = 0.119309875536981$

$\alpha_{51} = 0.048147179264990$ $\beta_{22} = 0.070605579799433$ $\beta_{51} = 0.010888081702583$

$\alpha_{52} = 0.151029729585865$ $\beta_{32} = 0.180764254304414$ $\beta_{52} = 0.034154109552284$

$\alpha_{54} = 0.800823091149145$ $\beta_{33} = 0.070606483961727$ $\beta_{54} = 0.181099440898861$

7.4.1.3 Coefficients of the optimal 5-stage 4th order implicit SSP RK

$\alpha_{21} = 1$

$\alpha_{32} = 0.785413771753555$ $\beta_{11} = 0.072154507748981$ $\beta_{44} = 0.077017601068238$

$\alpha_{43} = 0.934991917505507$ $\beta_{21} = 0.165562779595956$ $\beta_{54} = 0.158089969701175$

$\alpha_{54} = 0.954864191619538$ $\beta_{22} = 0.071232036614272$ $\beta_{55} = 0.106426690493882$

$\alpha_{65} = 0.894472670673021$ $\beta_{32} = 0.130035287184462$ $\beta_{65} = 0.148091381629243$

$\alpha_{52} = 0.045135808380468$ $\beta_{33} = 0.063186062090477$ $\beta_{52} = 0.007472809894781$

$\alpha_{62} = 0.105527329326976$ $\beta_{43} = 0.154799860761964$ $\beta_{62} = 0.017471397966712$

7.4.1.4 Coefficients of the optimal 6-stage 4th order implicit SSP RK

$\beta_{11} = 0.077219435861458$

$\beta_{21} = 0.128204308556198$ $\beta_{54} = 0.128204308556197$ $\alpha_{21} = \alpha_{32} = \alpha_{54} = 1$

$\beta_{22} = 0.063842903854499$ $\beta_{55} = 0.064105484788524$ $\alpha_{41} = 0.065974025631326$

$\beta_{32} = 0.128204308556197$ $\beta_{63} = 0.008043763906343$ $\alpha_{43} = 0.805203213502341$

$\beta_{33} = 0.058359965096908$ $\beta_{65} = 0.120160544649854$ $\alpha_{63} = 0.062741759593964$

$\beta_{41} = 0.008458154338733$ $\beta_{66} = 0.077016336936138$ $\alpha_{65} = 0.937258240406037$

$\beta_{43} = 0.103230521234296$ $\beta_{73} = 0.013804194371285$ $\alpha_{73} = 0.107673404480272$

$\beta_{44} = 0.058105933032597$ $\beta_{76} = 0.114400114184912$ $\alpha_{76} = 0.892326595519728$

7.4.1.5 Coefficients of the optimal 7-stage 4th order implicit SSP RK

$\beta_{11} = 0.081324471088377$ $\beta_{55} = 0.040474271914787$ $\alpha_{21} = 1$

$\beta_{21} = 0.108801609187400$ $\beta_{65} = 0.108801609187400$ $\alpha_{32} = 1$

$\beta_{22} = 0.051065224656204$ $\beta_{66} = 0.061352000212100$ $\alpha_{54} = \alpha_{65} = 1$

$\beta_{32} = 0.108801609187400$ $\beta_{73} = 0.020631403945188$ $\alpha_{43} = 0.865661994183934$

$\beta_{33} = 0.036491713577701$ $\beta_{76} = 0.088170205242212$ $\alpha_{73} = 0.189624069894518$

$\beta_{43} = 0.094185417979586$ $\beta_{77} = 0.080145231879588$ $\alpha_{76} = 0.810375930105481$

$\beta_{44} = 0.037028821732794$ $\beta_{83} = 0.001561606596621$ $\alpha_{83} = 0.014352789524754$

$\beta_{54} = 0.108801609187400$ $\beta_{87} = 0.107240002590779$ $\alpha_{87} = 0.985647210475246$

7.4.1.6 Coefficients of the optimal 8-stage 4th order implicit SSP RK

$\beta_{11} = 0.080355939553359$	$\beta_{55} = 0.030116385482588$	$\alpha_{21} = 1$
$\beta_{21} = 0.093742212796061$	$\beta_{65} = 0.093742212796061$	$\alpha_{32} = \alpha_{43} = 1$
$\beta_{22} = 0.054617345411549$	$\beta_{66} = 0.038334326442344$	$\alpha_{51} = 0.047220157287989$
$\beta_{32} = 0.093742212796061$	$\beta_{76} = 0.093742212796061$	$\alpha_{54} = 0.887270992114641$
$\beta_{33} = 0.039438131644116$	$\beta_{77} = 0.058861620081910$	$\alpha_{65} = 1$
$\beta_{43} = 0.093742212796061$	$\beta_{84} = 0.021977226754808$	$\alpha_{76} = 1$
$\beta_{44} = 0.032427875074076$	$\beta_{87} = 0.071764986041253$	$\alpha_{84} = 0.234443225728203$
$\beta_{51} = 0.004426522032754$	$\beta_{88} = 0.055606577879005$	$\alpha_{87} = 0.765556774271797$
$\beta_{54} = 0.083174746150582$	$\beta_{98} = 0.093742212796061$	$\alpha_{98} = 1$

7.4.1.7 Coefficients of the optimal 9-stage 4th order implicit SSP RK

$\beta_{11} = 0.068605696784244$		
$\beta_{21} = 0.082269487560004$	$\beta_{65} = 0.072971983212453$	$\alpha_{21} = 0.990643355064403$
$\beta_{22} = 0.048685583036902$	$\beta_{66} = 0.029699905991308$	$\alpha_{32} = 0.936520713898770$
$\beta_{32} = 0.077774790319743$	$\beta_{76} = 0.083046524401968$	$\alpha_{43} = \alpha_{54} = 1$
$\beta_{33} = 0.039925150083662$	$\beta_{77} = 0.035642110881905$	$\alpha_{61} = 0.105338196876962$
$\beta_{43} = 0.083046524401968$	$\beta_{87} = 0.083046524401969$	$\alpha_{62} = 0.015973817828813$
$\beta_{44} = 0.031928917146492$	$\beta_{88} = 0.050978240433952$	$\alpha_{65} = 0.878687985294225$
$\beta_{54} = 0.083046524401968$	$\beta_{95} = 0.017775897980583$	$\alpha_{76} = \alpha_{87} = 1$
$\beta_{55} = 0.029618614941264$	$\beta_{98} = 0.065270626421385$	$\alpha_{95} = 0.214047464461523$
$\beta_{61} = 0.008747971137402$	$\beta_{99} = 0.057552171403649$	$\alpha_{98} = 0.785952535538477$
$\beta_{62} = 0.001326570052113$	$\beta_{10,9} = 0.083046524401968$	$\alpha_{10,9} = 1$

7.4.1.8 Coefficients of the optimal 10-stage 4th order implicit SSP RK

$\beta_{11} = 0.053637857412307$	$\beta_{72} = 0.008896701400356$	$\alpha_{21} = 1$
$\beta_{21} = 0.073302847899924$	$\beta_{76} = 0.064406146499568$	$\alpha_{32} = 0.869472632481021$
$\beta_{22} = 0.042472343576273$	$\beta_{77} = 0.033369849008191$	$\alpha_{43} = 0.990280128291965$
$\beta_{32} = 0.063734820131903$	$\beta_{87} = 0.073302847899924$	$\alpha_{54} = \alpha_{65} = 1$
$\beta_{33} = 0.039816143518898$	$\beta_{88} = 0.037227578299133$	$\alpha_{72} = 0.121369109867354$
$\beta_{43} = 0.072590353622503$	$\beta_{98} = 0.073302847899924$	$\alpha_{76} = 0.878630890132646$
$\beta_{44} = 0.034233821696022$	$\beta_{99} = 0.046126339053885$	$\alpha_{87} = 1$
$\beta_{54} = 0.073302847899924$	$\beta_{10,6} = 0.012892211367605$	$\alpha_{98} = 1$
$\beta_{55} = 0.030626774272464$	$\beta_{10,9} = 0.060410636532319$	$\alpha_{10,6} = 0.175875995775857$
$\beta_{65} = 0.073302847899924$	$\beta_{10,10} = 0.053275700719583$	$\alpha_{10,9} = 0.824124004224143$
$\beta_{66} = 0.029485772863308$	$\beta_{11,10} = 0.073302847899924$	$\alpha_{11,10} = 1$

7.4.1.9   Coefficients of the optimal 11-stage 4th order implicit SSP RK

$\beta_{11} = 0.056977945207836$ $\beta_{76} = 0.065880156369595$ $\beta_{12,11} = 0.065880156369595$
$\beta_{21} = 0.065880156369595$ $\beta_{77} = 0.029602951078198$ $\alpha_{21} = \alpha_{32} = 1$
$\beta_{22} = 0.043484869703481$ $\beta_{83} = 0.009935800759662$ $\alpha_{41} = 0.000403688802047$
$\beta_{32} = 0.065880156369595$ $\beta_{87} = 0.055944355609932$ $\alpha_{43} = 0.929154313811668$
$\beta_{33} = 0.035790792116714$ $\beta_{88} = 0.027887296332663$ $\alpha_{54} = \alpha_{65} = \alpha_{76} = 1$
$\beta_{41} = 0.000026595081404$ $\beta_{98} = 0.065880156369595$ $\alpha_{83} = 0.150816289869158$
$\beta_{43} = 0.061212831485396$ $\beta_{99} = 0.033340440672342$ $\alpha_{87} = 0.849183710130842$
$\beta_{44} = 0.029306212740362$ $\beta_{10,9} = 0.065880156369595$ $\alpha_{98} = 1$
$\beta_{54} = 0.065880156369595$ $\beta_{10,10} = 0.042024506703707$ $\alpha_{10,9} = 1$
$\beta_{55} = 0.028274789742965$ $\beta_{11,7} = 0.012021727578515$ $\alpha_{11,7} = 0.182478734735714$
$\beta_{65} = 0.065880156369595$ $\beta_{11,10} = 0.053858428791080$ $\alpha_{11,10} = 0.817521265264286$
$\beta_{66} = 0.025442782369057$ $\beta_{11,11} = 0.045164424313434$ $\alpha_{12,11} = 1$

7.4.2   Fifth order methods

7.4.2.1   Coefficients of the optimal 4-stage 5th order implicit SSP RK

$\beta_{21} = 0.125534208080981$ $\beta_{44} = 0.133639210602434$ $\alpha_{41} = 0.111760167014216$
$\beta_{22} = 0.125534208080983$ $\beta_{51} = 0.022869941925234$ $\alpha_{42} = 0.000000006110058$
$\beta_{32} = 0.350653119567098$ $\beta_{52} = 0.138100556728488$ $\alpha_{43} = 0.462033126016285$
$\beta_{33} = 0.048181647388277$ $\beta_{53} = 0.157510964003014$ $\alpha_{51} = 0.026143376902960$
$\beta_{41} = 0.097766579224131$ $\beta_{54} = 0.277310825799681$ $\alpha_{52} = 0.157867252871240$
$\beta_{42} = 0.000000005345013$ $\alpha_{21} = 0.143502249669229$ $\alpha_{53} = 0.180055922824003$
$\beta_{43} = 0.404181556145118$ $\alpha_{32} = 0.400843023432714$ $\alpha_{54} = 0.317003054133379$

7.4.2.2   Coefficients of the optimal 5-stage 5th order implicit SSP RK

$\beta_{21} = 0.107733237609082$ $\beta_{52} = 0.011356303341111$ $\alpha_{41} = 0.035170229692428$
$\beta_{22} = 0.107733237609079$ $\beta_{53} = 0.024232322953809$ $\alpha_{42} = 0.000000100208717$
$\beta_{31} = 0.000009733684024$ $\beta_{54} = 0.220980752503271$ $\alpha_{43} = 0.786247596634378$
$\beta_{32} = 0.205965878618791$ $\beta_{55} = 0.098999612937858$ $\alpha_{51} = 0.128913001605754$
$\beta_{33} = 0.041505157180052$ $\beta_{63} = 0.079788022937926$ $\alpha_{52} = 0.036331447472278$
$\beta_{41} = 0.010993335656900$ $\beta_{64} = 0.023678103998428$ $\alpha_{53} = 0.077524819660326$
$\beta_{42} = 0.000000031322743$ $\beta_{65} = 0.194911604040485$ $\alpha_{54} = 0.706968664080396$
$\beta_{43} = 0.245761367350216$ $\alpha_{21} = 0.344663606249694$ $\alpha_{63} = 0.255260385110718$
$\beta_{44} = 0.079032059834967$ $\alpha_{31} = 0.000031140312055$ $\alpha_{64} = 0.075751744720289$
$\beta_{51} = 0.040294985548405$ $\alpha_{32} = 0.658932601159987$ $\alpha_{65} = 0.623567413728619$

7.4.2.3 Coefficients of the optimal 6-stage 5th order implicit SSP RK

$\beta_{21} = 0.084842972180459$

$\beta_{22} = 0.084842972180464$ $\beta_{65} = 0.159145416202648$ $\alpha_{62} = 0.072495338903420$

$\beta_{32} = 0.149945333907731$ $\beta_{66} = 0.085074359110886$ $\alpha_{63} = 0.133329934574294$

$\beta_{33} = 0.063973483119994$ $\beta_{73} = 0.004848530454093$ $\alpha_{65} = 0.791612404723054$

$\beta_{43} = 0.175767531234932$ $\beta_{74} = 0.042600565019890$ $\alpha_{73} = 0.024117294382203$

$\beta_{44} = 0.055745328618053$ $\beta_{76} = 0.151355691945479$ $\alpha_{74} = 0.211901395105308$

$\beta_{51} = 0.024709139041008$ $\alpha_{21} = 0.422021261021445$ $\alpha_{76} = 0.752865185365536$

$\beta_{54} = 0.173241563951140$ $\alpha_{32} = 0.745849859731775$ $\alpha_{65} = 0.791612404723054$

$\beta_{55} = 0.054767418942828$ $\alpha_{43} = 0.874293218071360$ $\alpha_{73} = 0.024117294382203$

$\beta_{62} = 0.014574431645716$ $\alpha_{51} = 0.122906844831659$ $\alpha_{74} = 0.211901395105308$

$\beta_{63} = 0.026804592504486$ $\alpha_{54} = 0.861728690085026$ $\alpha_{76} = 0.752865185365536$

7.4.2.4 Coefficients of the optimal 7-stage 5th order implicit SSP RK

$\beta_{21} = 0.077756487471956$ $\beta_{66} = 0.037306165750735$

$\beta_{22} = 0.077756487471823$ $\beta_{73} = 0.020177924440034$

$\beta_{32} = 0.126469010941083$ $\beta_{76} = 0.140855998083160$ $\alpha_{54} = 0.900717090387559$

$\beta_{33} = 0.058945597921853$ $\beta_{77} = 0.077972159279168$ $\alpha_{62} = 0.071128941372444$

$\beta_{43} = 0.143639250502198$ $\beta_{84} = 0.009653207936821$ $\alpha_{63} = 0.168525096484428$

$\beta_{44} = 0.044443238891736$ $\beta_{85} = 0.025430639631870$ $\alpha_{65} = 0.760345962143127$

$\beta_{51} = 0.011999093244164$ $\beta_{86} = 0.000177781270869$ $\alpha_{73} = 0.125302322168346$

$\beta_{54} = 0.145046006148787$ $\beta_{87} = 0.124996366168017$ $\alpha_{76} = 0.874697677831654$

$\beta_{55} = 0.047108760907057$ $\alpha_{21} = 0.482857811904546$ $\alpha_{84} = 0.059945182887979$

$\beta_{62} = 0.011454172434127$ $\alpha_{32} = 0.785356333370487$ $\alpha_{85} = 0.157921009644458$

$\beta_{63} = 0.027138257330487$ $\alpha_{43} = 0.891981318293413$ $\alpha_{86} = 0.001103998884730$

$\beta_{65} = 0.122441492758580$ $\alpha_{51} = 0.074512829695468$ $\alpha_{87} = 0.776211398253764$

7.4.2.5 Coefficients of the optimal 8-stage 5th order implicit SSP RK

$\beta_{21} = 0.068228425119547$ $\beta_{73} = 0.015335646668415$ $\alpha_{51} = 0.069260513476804$

$\beta_{22} = 0.068228425081188$ $\beta_{76} = 0.116977452926909$ $\alpha_{54} = 0.908882077064212$

$\beta_{32} = 0.105785458668142$ $\beta_{77} = 0.050447703819928$ $\alpha_{62} = 0.056144626483417$

$\beta_{33} = 0.049168429086829$ $\beta_{84} = 0.011255581082016$ $\alpha_{63} = 0.148913610539984$

$\beta_{43} = 0.119135238085849$ $\beta_{85} = 0.006541409424671$ $\alpha_{65} = 0.794939486396848$

$\beta_{44} = 0.040919294063196$ $\beta_{87} = 0.114515518273119$ $\alpha_{73} = 0.115904148048060$

$\beta_{51} = 0.009164078944895$ $\beta_{88} = 0.060382824328534$ $\alpha_{76} = 0.884095226988328$

$\beta_{54} = 0.120257079939301$ $\beta_{95} = 0.002607774587593$ $\alpha_{84} = 0.085067722561958$

$\beta_{55} = 0.039406904101415$ $\beta_{96} = 0.024666705635997$ $\alpha_{85} = 0.049438833770315$

$\beta_{62} = 0.007428674198294$ $\beta_{98} = 0.104666894951906$ $\alpha_{87} = 0.865488353423280$

$\beta_{63} = 0.019703233696280$ $\alpha_{21} = 0.515658560550227$ $\alpha_{95} = 0.019709106398420$

$\beta_{65} = 0.105180973170163$ $\alpha_{32} = 0.799508082567950$ $\alpha_{96} = 0.186426667470161$

$\beta_{66} = 0.045239659320409$ $\alpha_{43} = 0.900403391614526$ $\alpha_{98} = 0.791054172708715$

7.4.2.6 Coefficients of the optimal 9-stage 5th order implicit SSP RK

$\beta_{21} = 0.057541273792734$

$\beta_{22} = 0.057541282875429$

$\beta_{32} = 0.089687860942851$

$\beta_{33} = 0.041684970395150$

$\beta_{43} = 0.101622955619526$

$\beta_{44} = 0.040743690263377$

$\beta_{51} = 0.009276188714858$

$\beta_{54} = 0.101958242208571$

$\beta_{55} = 0.040815264589441$

$\beta_{62} = 0.011272987717036$

$\beta_{65} = 0.101125244372555$

$\beta_{66} = 0.040395338505384$

$\beta_{73} = 0.003606182878823$

$\beta_{74} = 0.018205434656765$

$\beta_{76} = 0.090586614534056$

$\beta_{77} = 0.042925976445877$

$\beta_{84} = 0.011070977346914$

$\beta_{87} = 0.101327254746568$

$\beta_{88} = 0.046669302312152$

$\beta_{95} = 0.010281040119047$

$\beta_{98} = 0.102117191974435$

$\beta_{99} = 0.050500143250113$

$\beta_{10,6} = 0.000157554758807$

$\beta_{10,7} = 0.023607648002010$

$\beta_{10,9} = 0.088454624345414$

$\alpha_{21} = 0.511941093031398$

$\alpha_{32} = 0.797947256574797$

$\alpha_{43} = 0.904133043080300$

$\alpha_{51} = 0.082529667434119$

$\alpha_{54} = 0.907116066770269$

$\alpha_{62} = 0.100295062538531$

$\alpha_{65} = 0.899704937426848$

$\alpha_{73} = 0.032083982209117$

$\alpha_{74} = 0.161972606843345$

$\alpha_{76} = 0.805943410735452$

$\alpha_{84} = 0.098497788983963$

$\alpha_{87} = 0.901502211016037$

$\alpha_{95} = 0.091469767162319$

$\alpha_{98} = 0.908530232837680$

$\alpha_{10,6} = 0.001401754777391$

$\alpha_{10,7} = 0.210035759124536$

$\alpha_{10,9} = 0.786975228149903$

7.4.2.7 Coefficients of the optimal 10-stage 5th order implicit SSP RK

$\beta_{84} = 0.009638972523544$

$\beta_{21} = 0.052445615058994$

$\beta_{22} = 0.052445635165954$

$\beta_{32} = 0.079936220395519$

$\beta_{33} = 0.038724845476313$

$\beta_{43} = 0.089893189589075$

$\beta_{44} = 0.037676214671832$

$\beta_{51} = 0.007606429497294$

$\beta_{54} = 0.090180506502554$

$\beta_{55} = 0.035536573874530$

$\beta_{62} = 0.009295158915663$

$\beta_{65} = 0.089447242753894$

$\beta_{66} = 0.036490114423762$

$\beta_{73} = 0.003271387942850$

$\beta_{74} = 0.015255382390056$

$\beta_{76} = 0.080215515252923$

$\beta_{77} = 0.035768398609662$

$\beta_{87} = 0.089103469454345$

$\beta_{88} = 0.040785658461768$

$\beta_{95} = 0.009201462517982$

$\beta_{98} = 0.089540979697808$

$\beta_{99} = 0.042414168555682$

$\beta_{10,6} = 0.005634796609556$

$\beta_{10,7} = 0.006560464576444$

$\beta_{10,9} = 0.086547180546464$

$\beta_{10,10} = 0.043749770437420$

$\beta_{11,7} = 0.001872759401284$

$\beta_{11,8} = 0.017616881402665$

$\beta_{11,10} = 0.079160150775900$

$\alpha_{21} = 0.531135486241871$

$\alpha_{32} = 0.809542670828687$

$\alpha_{43} = 0.910380456183399$

$\alpha_{51} = 0.077033029836054$

$\alpha_{54} = 0.913290217244921$

$\alpha_{62} = 0.094135396158718$

$\alpha_{65} = 0.905864193215084$

$\alpha_{73} = 0.033130514796271$

$\alpha_{74} = 0.154496709294644$

$\alpha_{76} = 0.812371189661489$

$\alpha_{84} = 0.097617319434729$

$\alpha_{87} = 0.902382678155958$

$\alpha_{95} = 0.093186499255038$

$\alpha_{98} = 0.906813500744962$

$\alpha_{10,6} = 0.057065598977612$

$\alpha_{10,7} = 0.066440169285130$

$\alpha_{10,9} = 0.876494226842443$

$\alpha_{11,7} = 0.018966103726616$

$\alpha_{11,8} = 0.178412453726484$

$\alpha_{11,10} = 0.801683136446066$

7.4.2.8 Coefficients of the optimal 11-stage 5th order implicit SSP RK

$\beta_{21} = 0.048856948431570$

$\beta_{22} = 0.048856861697775$

$\beta_{32} = 0.072383163641108$

$\beta_{33} = 0.035920513887793$

$\beta_{43} = 0.080721632683704$

$\beta_{44} = 0.034009594943671$

$\beta_{51} = 0.006438090160799$

$\beta_{54} = 0.081035022899306$

$\beta_{55} = 0.032672027896742$

$\beta_{62} = 0.007591099341932$

$\beta_{63} = 0.000719846382100$

$\beta_{65} = 0.079926841108108$

$\beta_{66} = 0.033437798720082$

$\beta_{73} = 0.003028997848550$

$\beta_{74} = 0.012192534706212$

$\beta_{76} = 0.073016254277378$

$\beta_{77} = 0.033377699686911$

$\beta_{84} = 0.008251011235053$

$\beta_{87} = 0.079986775597087$

$\beta_{88} = 0.035640440183022$

$\beta_{95} = 0.008095394925904$

$\beta_{98} = 0.080142391870059$

$\beta_{99} = 0.036372965664654$

$\alpha_{21} = 0.553696439876870$

$\alpha_{32} = 0.820319346617409$

$\alpha_{43} = 0.914819326070196$

$\alpha_{51} = 0.072962960562995$

$\alpha_{54} = 0.918370981510030$

$\alpha_{62} = 0.086030028794504$

$\alpha_{63} = 0.008158028526592$

$\alpha_{65} = 0.905811942678904$

$\alpha_{73} = 0.034327672500586$

$\alpha_{74} = 0.138178156365216$

$\alpha_{76} = 0.827494171134198$

$\alpha_{84} = 0.093508818968334$

$\alpha_{87} = 0.906491181031666$

$\alpha_{95} = 0.091745217287743$

$\alpha_{98} = 0.908254782302260$

$\beta_{10,6} = 0.005907318148947$

$\beta_{10,7} = 0.005394911565057$

$\beta_{10,9} = 0.076935557118137$

$\beta_{10,10} = 0.032282094274356$

$\beta_{11,7} = 0.003571080721480$

$\beta_{11,8} = 0.008920593887617$

$\beta_{11,10} = 0.075746112223043$

$\beta_{11,11} = 0.042478561828713$

$\beta_{12,8} = 0.004170617993886$

$\beta_{12,9} = 0.011637432775226$

$\beta_{12,11} = 0.072377330912325$

$\alpha_{10,6} = 0.066947714363965$

$\alpha_{10,7} = 0.061140603801867$

$\alpha_{10,9} = 0.871911681834169$

$\alpha_{11,7} = 0.040471104837131$

$\alpha_{11,8} = 0.101097207986272$

$\alpha_{11,10} = 0.858431687176596$

$\alpha_{12,8} = 0.047265668639449$

$\alpha_{12,9} = 0.131887178872293$

$\alpha_{12,11} = 0.820253244225314$

7.4.3 Sixth order methods

7.4.3.1 Coefficients of the optimal 6-stage 6th order implicit SSP RK

$\beta_{21} = 0.306709397198437$

$\beta_{22} = 0.306709397198281$

$\beta_{31} = 0.100402778173265$

$\beta_{32} = 0.000000014622272$

$\beta_{33} = 0.100402700098726$

$\beta_{41} = 0.000015431349319$

$\beta_{42} = 0.000708584139276$

$\beta_{43} = 0.383195003696784$

$\beta_{44} = 0.028228318307509$

$\beta_{51} = 0.101933808745384$

$\beta_{52} = 0.000026687930165$

$\beta_{53} = 0.136711477475771$

$\beta_{54} = 0.331296656179688$

$\beta_{55} = 0.107322255666019$

$\beta_{61} = 0.000033015066992$

$\beta_{62} = 0.000000017576816$

$\beta_{63} = 0.395057247524893$

$\beta_{64} = 0.014536993458566$

$\beta_{65} = 0.421912313467517$

$\beta_{66} = 0.049194928995335$

$\beta_{71} = 0.054129307323559$

$\beta_{72} = 0.002083586568620$

$\beta_{73} = 0.233976271277479$

$\beta_{74} = 0.184897163424393$

$\beta_{75} = 0.303060566272042$

$\beta_{76} = 0.135975816243004$

$\alpha_{21} = 0.055928810359256$

$\alpha_{31} = 0.018308561756789$

$\alpha_{32} = 0.000000002666388$

$\alpha_{41} = 0.000002813924247$

$\alpha_{42} = 0.000129211130507$

$\alpha_{43} = 0.069876048429340$

$\alpha_{51} = 0.018587746937629$

$\alpha_{52} = 0.000004866574675$

$\alpha_{53} = 0.024929494718837$

$\alpha_{54} = 0.060412325234826$

$\alpha_{61} = 0.000006020335333$

$\alpha_{62} = 0.000000003205153$

$\alpha_{63} = 0.072039142196788$

$\alpha_{64} = 0.002650837430364$

$\alpha_{65} = 0.076936194272824$

$\alpha_{71} = 0.009870541274021$

$\alpha_{72} = 0.000379944400556$

$\alpha_{73} = 0.042665841426363$

$\alpha_{74} = 0.033716209818106$

$\alpha_{75} = 0.055263441854804$

$\alpha_{76} = 0.024795346049276$

7.4.3.2   Coefficients of the optimal 7-stage 6th order implicit SSP RK

$\beta_{21} = 0.090485932570398$

$\beta_{22} = 0.090485932570397$

$\beta_{32} = 0.346199513509666$

$\beta_{33} = 0.056955495796615$

$\beta_{41} = 0.089183260058590$

$\beta_{42} = 0.122181527536711$

$\beta_{43} = 0.340520235772773$

$\beta_{44} = 0.086699362107543$

$\beta_{51} = 0.214371998459638$

$\beta_{52} = 0.046209156887254$

$\beta_{53} = 0.215162143673919$

$\beta_{54} = 0.000000362542364$

$\beta_{55} = 0.209813410800754$

$\beta_{61} = 0.000000591802702$

$\beta_{62} = 0.390556634551239$

$\beta_{63} = 0.000000491944026$

$\beta_{64} = 0.330590135449081$

$\beta_{65} = 0.007410530577593$

$\beta_{66} = 0.070407008959133$

$\beta_{71} = 0.000000021842570$

$\beta_{72} = 0.325421794191472$

$\beta_{73} = 0.069025907032937$

$\beta_{74} = 0.373360315300742$

$\beta_{75} = 0.007542750523234$

$\beta_{76} = 0.005465714557738$

$\beta_{77} = 0.063240270982556$

$\beta_{81} = 0.044161355044152$

$\beta_{82} = 0.204837996136028$

$\beta_{83} = 0.191269829083813$

$\beta_{84} = 0.255834644704751$

$\beta_{85} = 0.015984178241749$

$\beta_{86} = 0.016124165979879$

$\beta_{87} = 0.151145768228502$

$\alpha_{21} = 0.023787133610744$

$\alpha_{32} = 0.091009661390427$

$\alpha_{41} = 0.023444684301672$

$\alpha_{42} = 0.032119338749362$

$\alpha_{43} = 0.089516680829776$

$\alpha_{51} = 0.056354565012571$

$\alpha_{52} = 0.012147561037311$

$\alpha_{53} = 0.056562280060094$

$\alpha_{54} = 0.000000095305905$

$\alpha_{61} = 0.000000155574348$

$\alpha_{62} = 0.102670355321862$

$\alpha_{63} = 0.000000129323288$

$\alpha_{64} = 0.086906235023916$

$\alpha_{65} = 0.001948095974350$

$\alpha_{71} = 0.000000005742021$

$\alpha_{72} = 0.085547570527144$

$\alpha_{73} = 0.018145676643359$

$\alpha_{74} = 0.098149750494075$

$\alpha_{75} = 0.001982854233713$

$\alpha_{76} = 0.001436838619770$

$\alpha_{81} = 0.011609230551384$

$\alpha_{82} = 0.053848246287940$

$\alpha_{83} = 0.050281417794762$

$\alpha_{84} = 0.067254353278777$

$\alpha_{85} = 0.004201954631994$

$\alpha_{86} = 0.004238754905099$

$\alpha_{87} = 0.039733519691061$

7.4.3.3 Coefficients of the optimal 8-stage 6th order implicit SSP RK

$\beta_{21} = 0.078064586430339$

$\beta_{22} = 0.078064586430334$

$\beta_{31} = 0.000000000128683$

$\beta_{32} = 0.207887720440412$

$\beta_{33} = 0.051491724905522$

$\beta_{41} = 0.039407945831803$

$\beta_{43} = 0.256652317630585$

$\beta_{44} = 0.062490509654886$

$\beta_{51} = 0.009678931461971$

$\beta_{52} = 0.113739188386853$

$\beta_{54} = 0.227795405648863$

$\beta_{55} = 0.076375614721986$

$\beta_{62} = 0.010220279377975$

$\beta_{63} = 0.135083590682973$

$\beta_{65} = 0.235156310567507$

$\beta_{66} = 0.033370798931382$

$\beta_{72} = 0.000000009428737$

$\beta_{73} = 0.112827524882246$

$\beta_{74} = 0.001997541632150$

$\beta_{75} = 0.177750742549303$

$\beta_{76} = 0.099344022703332$

$\beta_{77} = 0.025183595544641$

$\beta_{81} = 0.122181071065616$

$\beta_{82} = 0.000859535946343$

$\beta_{83} = 0.008253954430873$

$\beta_{84} = 0.230190271515289$

$\beta_{85} = 0.046429529676480$

$\beta_{86} = 0.017457063072040$

$\beta_{87} = 0.017932893410781$

$\beta_{88} = 0.322331010725841$

$\beta_{91} = 0.011069087473717$

$\beta_{92} = 0.010971589676607$

$\beta_{93} = 0.068827453812950$

$\beta_{94} = 0.048864283062331$

$\beta_{95} = 0.137398274895655$

$\beta_{96} = 0.090347431612516$

$\beta_{97} = 0.029504401738350$

$\beta_{98} = 0.000167109498102$

$\alpha_{21} = 0.175964293749273$

$\alpha_{31} = 0.000000000290062$

$\alpha_{32} = 0.468596806556916$

$\alpha_{41} = 0.088828900190110$

$\alpha_{43} = 0.578516403866171$

$\alpha_{51} = 0.021817144198582$

$\alpha_{52} = 0.256377915663045$

$\alpha_{54} = 0.513470441684846$

$\alpha_{62} = 0.023037388973687$

$\alpha_{63} = 0.304490034708070$

$\alpha_{65} = 0.530062554633790$

$\alpha_{72} = 0.000000021253185$

$\alpha_{73} = 0.254322947692795$

$\alpha_{74} = 0.004502630688369$

$\alpha_{75} = 0.400665465691124$

$\alpha_{76} = 0.223929973789109$

$\alpha_{81} = 0.275406645480353$

$\alpha_{82} = 0.001937467969363$

$\alpha_{83} = 0.018605123379003$

$\alpha_{84} = 0.518868675379274$

$\alpha_{85} = 0.104656154246370$

$\alpha_{86} = 0.039349722004217$

$\alpha_{87} = 0.040422284523661$

$\alpha_{91} = 0.024950675444873$

$\alpha_{92} = 0.024730907022402$

$\alpha_{93} = 0.155143002154553$

$\alpha_{94} = 0.110144297841125$

$\alpha_{95} = 0.309707532056893$

$\alpha_{96} = 0.203650883489192$

$\alpha_{97} = 0.066505459796630$

$\alpha_{98} = 0.000376679185235$

7.4.3.4 Coefficients of the optimal 9-stage 6th order implicit SSP RK

$\beta_{21} = 0.060383920365295$

$\beta_{22} = 0.060383920365140$

$\beta_{31} = 0.000000016362287$

$\beta_{32} = 0.119393671070984$

$\beta_{33} = 0.047601859039825$

$\beta_{42} = 0.000000124502898$

$\beta_{43} = 0.144150297305350$

$\beta_{44} = 0.016490678866732$

$\beta_{51} = 0.014942049029658$

$\beta_{52} = 0.033143125204828$

$\beta_{53} = 0.020040368468312$

$\beta_{54} = 0.095855615754989$

$\beta_{55} = 0.053193337903908$

$\beta_{61} = 0.000006536159050$

$\beta_{62} = 0.000805531139166$

$\beta_{63} = 0.015191136635430$

$\beta_{64} = 0.054834245267704$

$\beta_{65} = 0.089706774214904$

$\beta_{71} = 0.000006097150226$

$\beta_{72} = 0.018675155382709$

$\beta_{73} = 0.025989306353490$

$\beta_{74} = 0.000224116890218$

$\beta_{75} = 0.000125522781582$

$\beta_{76} = 0.125570620920810$

$\beta_{77} = 0.019840674620006$

$\beta_{81} = 0.000000149127775$

$\beta_{82} = 0.000000015972341$

$\beta_{83} = 0.034242827620807$

$\beta_{84} = 0.017165973521939$

$\beta_{85} = 0.000000000381532$

$\beta_{86} = 0.001237807078917$

$\beta_{87} = 0.119875131948576$

$\beta_{88} = 0.056749019092783$

$\beta_{91} = 0.000000072610411$

$\beta_{92} = 0.000000387168511$

$\beta_{93} = 0.000400376164405$

$\beta_{94} = 0.000109472445726$

$\beta_{95} = 0.012817181286633$

$\beta_{96} = 0.011531979169562$

$\beta_{97} = 0.000028859233948$

$\beta_{98} = 0.143963789161172$

$\beta_{99} = 0.060174596046625$

$\alpha_{21} = 0.350007201986739$

$\alpha_{31} = 0.000000094841777$

$\alpha_{32} = 0.692049215977999$

$\alpha_{42} = 0.000000721664155$

$\alpha_{43} = 0.835547641163090$

$\alpha_{51} = 0.086609559981880$

$\alpha_{52} = 0.192109628653810$

$\alpha_{53} = 0.116161276908552$

$\alpha_{54} = 0.555614071795216$

$\alpha_{61} = 0.000037885959162$

$\alpha_{62} = 0.004669151960107$

$\alpha_{63} = 0.088053362494510$

$\alpha_{64} = 0.317839263219390$

$\alpha_{65} = 0.519973146034093$

$\alpha_{71} = 0.000035341304071$

$\alpha_{72} = 0.108248004479122$

$\alpha_{73} = 0.150643488255346$

$\alpha_{74} = 0.001299063147749$

$\alpha_{75} = 0.000727575773504$

$\alpha_{76} = 0.727853067743022$

$\alpha_{81} = 0.000000864398917$

$\alpha_{82} = 0.000000092581509$

$\alpha_{83} = 0.198483904509141$

$\alpha_{84} = 0.099500236576982$

$\alpha_{85} = 0.000000002211499$

$\alpha_{86} = 0.007174780797111$

$\alpha_{87} = 0.694839938634174$

$\alpha_{91} = 0.000000420876394$

$\alpha_{92} = 0.000002244169749$

$\alpha_{93} = 0.002320726117116$

$\alpha_{94} = 0.000634542179300$

$\alpha_{95} = 0.074293052394615$

$\alpha_{96} = 0.066843552689032$

$\alpha_{97} = 0.000167278634186$

$\alpha_{98} = 0.834466572009306$

$\beta_{10,1} = 0.001577092080021$

$\beta_{10,2} = 0.000008909587678$

$\beta_{10,3} = 0.000003226074427$

$\beta_{10,4} = 0.000000062166910$

$\beta_{10,5} = 0.009112668630420$

$\beta_{10,6} = 0.008694079174358$

$\beta_{10,7} = 0.017872872156132$

$\beta_{10,8} = 0.027432316305282$

$\beta_{10,9} = 0.107685980331284$

$\alpha_{10,1} = 0.009141400274516$

$\alpha_{10,2} = 0.000051643216195$

$\alpha_{10,3} = 0.000018699502726$

$\alpha_{10,4} = 0.000000360342058$

$\alpha_{10,5} = 0.052820347381733$

$\alpha_{10,6} = 0.050394050390558$

$\alpha_{10,7} = 0.103597678603687$

$\alpha_{10,8} = 0.159007699664781$

$\alpha_{10,9} = 0.624187175011814$

7.4.3.5 Coefficients of the optimal 10-stage 6th order implicit SSP RK

$\alpha_{96} = 0.099613661566658$

$\beta_{21} = 0.054638144097621$

$\beta_{22} = 0.054638144097609$

$\beta_{32} = 0.094708145223810$

$\beta_{33} = 0.044846931722606$

$\beta_{43} = 0.108958403164940$

$\beta_{44} = 0.031071352647397$

$\beta_{51} = 0.004498251069701$

$\beta_{52} = 0.005530448043688$

$\beta_{54} = 0.107851443619437$

$\beta_{55} = 0.018486380725450$

$\beta_{62} = 0.015328210231111$

$\beta_{63} = 0.014873940010974$

$\beta_{64} = 0.000000013999299$

$\beta_{65} = 0.093285690103096$

$\beta_{66} = 0.031019852663844$

$\beta_{73} = 0.023345108682580$

$\beta_{74} = 0.000000462051194$

$\beta_{76} = 0.100142283610706$

$\beta_{77} = 0.037191650574052$

$\beta_{84} = 0.020931607249912$

$\beta_{85} = 0.007491225374492$

$\beta_{86} = 0.000000004705702$

$\beta_{87} = 0.094887152674486$

$\beta_{88} = 0.041052752299292$

$\beta_{94} = 0.000000000437894$

$\beta_{95} = 0.013484714992727$

$\beta_{96} = 0.012301077330264$

$\beta_{98} = 0.097178530400423$

$\beta_{99} = 0.039273658398104$

$\alpha_{21} = 0.442457635916190$

$\alpha_{32} = 0.766942997969774$

$\alpha_{43} = 0.882341050812911$

$\alpha_{51} = 0.036426667979449$

$\alpha_{52} = 0.044785360253007$

$\alpha_{54} = 0.873376934047102$

$\alpha_{62} = 0.124127269944714$

$\alpha_{63} = 0.120448606787528$

$\alpha_{64} = 0.000000113365798$

$\alpha_{65} = 0.755424009901960$

$\alpha_{73} = 0.189047812082446$

$\alpha_{74} = 0.000003741673193$

$\alpha_{76} = 0.810948446244362$

$\alpha_{84} = 0.169503368254511$

$\alpha_{85} = 0.060663661331375$

$\alpha_{86} = 0.000000038106595$

$\alpha_{87} = 0.768392593572726$

$\alpha_{94} = 0.000000003546047$

$\alpha_{95} = 0.109198714839684$

$\alpha_{98} = 0.786948084216732$

$\beta_{10,1} = 0.000987065715240$

$\beta_{10,2} = 0.000000347467847$

$\beta_{10,6} = 0.004337021151393$

$\beta_{10,7} = 0.011460261685365$

$\beta_{10,8} = 0.002121689510807$

$\beta_{10,9} = 0.104338127248348$

$\beta_{10,10} = 0.042268075457472$

$\beta_{11,3} = 0.000656941338471$

$\beta_{11,7} = 0.015039465910057$

$\beta_{11,8} = 0.004816543620956$

$\beta_{11,9} = 0.031302441038151$

$\beta_{11,10} = 0.071672462436845$

$\alpha_{10,1} = 0.007993221037648$

$\alpha_{10,2} = 0.000002813781560$

$\alpha_{10,6} = 0.035121034164983$

$\alpha_{10,7} = 0.092804768098049$

$\alpha_{10,8} = 0.017181361859997$

$\alpha_{10,9} = 0.844926230212794$

$\alpha_{11,3} = 0.005319886250823$

$\alpha_{11,7} = 0.121789029292733$

$\alpha_{11,8} = 0.039004189088262$

$\alpha_{11,9} = 0.253485990215933$

$\alpha_{11,10} = 0.580400905152248$

Chapter 8

SSP Properties of Linear Multistep Methods

Up to now, we have considered primarily the SSP properties of Runge–Kutta methods. Linear multistep methods can also be written as convex combinations of forward Euler methods, so that they, too, are SSP. In this chapter, we consider the SSP properties of linear multistep methods.

Unlike Runge–Kutta methods, which have multiple representations, explicit s step methods have the unique form [91],

$$u^{n+1} = \sum_{i=1}^{s} \left(\alpha_i u^{n+1-i} + \Delta t \beta_i F(u^{n+1-i}) \right),$$

$$= \sum_{i=1}^{s} \alpha_i \left(u^{n+1-i} + \Delta t \frac{\beta_i}{\alpha_i} F(u^{n+1-i}) \right), \qquad (8.1)$$

consistency requires that

$$\sum_{i=1}^{s} \alpha_i = 1. \qquad (8.2)$$

(For implicit methods, we allow the summation to go from zero.) As usual, we consider initial value ODEs $u' = F(u)$ where F is such that the forward Euler method is strongly stable

$$\|u + \Delta t F(u)\| \leq \|u\| \text{ for } 0 \leq \Delta t \leq \Delta t_{\text{FE}} \text{ for all } u. \qquad (8.3)$$

Then (8.1) and (8.2) imply that, if the coefficients α_i and β_i are nonnegative, this multistep method is SSP:

$$\|u^{n+1}\| \leq \max \left\{ \|u^n\|, \|u^{n-1}\|, \ldots, \|u^{n-s+1}\| \right\} \qquad (8.4)$$

for time steps satisfying

$$\Delta t \leq \min_i \frac{\alpha_i}{\beta_i} \Delta t_{FE}.$$

Thus the SSP coefficient of the multistep method (8.1) is simply

$$\mathcal{C} = \begin{cases} \min \frac{\alpha_i}{\beta_i} & \text{if } \alpha_i, \beta_i \geq 0 \text{ for all } i \\ 0 & \text{otherwise.} \end{cases}$$

As usual, the ratio above is taken to be infinite when $\beta_i = 0$.

The order conditions can be easily derived using Taylor expansions by substituting the exact solution U at times $t_n = t_0 + n\Delta t$ in the numerical scheme (8.1):

$$U(t_{n+1}) - \sum_{i=1}^{s} \left(\alpha_i U(t_{n+1-i}) - \Delta t \beta_i F(U(t_{n+1-i})) \right)$$

$$= \sum_{i=1}^{s} \alpha_i U(t_{n+1}) - \sum_{i=1}^{s} \left(\alpha_i U(t_{n+1-i}) - \Delta t \beta_i F(U(t_{n+1-i})) \right)$$

$$= \sum_{i=1}^{s} \alpha_i U(t_{n+1}) - \sum_{i=1}^{s} \alpha_i \sum_{k=0}^{p} \frac{1}{k!} (-i\Delta t)^k U^{(k)}(t_{n+1})$$

$$- \Delta t \sum_{i=1}^{s} \beta_i \sum_{k=0}^{p-1} \frac{1}{k!} (-i\Delta t)^k U^{(k+1)}(t_{n+1})$$

$$= - \sum_{k=1}^{p} (\Delta t)^k \sum_{i=1}^{s} \left(\alpha_i \frac{1}{k!} (-i)^k + \beta_i \frac{1}{(k-1)!} (-i)^{k-1} \right) U^{(k)}(t_{n+1}).$$

Thus, for a method of the form (8.1) to be pth order, its coefficients must satisfy the consistency condition (8.2) and the p order conditions

$$\sum_{i=1}^{s} i^k \alpha_i = k \left(\sum_{i=1}^{s} i^{k-1} \beta_i \right) \qquad k = 1, ..., p. \tag{8.5}$$

Unlike Runge–Kutta methods, linear multistep methods are subject to the same order conditions whether they are applied to linear or nonlinear problems. An advantage of SSP multistep methods is that they do not suffer from the order barrier that SSP Runge–Kutta methods are subject to. Thus, we focus on finding multistep methods that are optimal in the sense of allowable time step. As we will see below, many of the optimal multistep methods have the same efficiency as the optimal Runge–Kutta methods for the same order of accuracy. In this chapter, we only consider the situation that both α_i's and β_i's in Equation (8.1) are non-negative. The general situation with possible negative β_i's will be considered in Chapter 10.

In practice, SSP multistep methods have been less often used than SSP Runge–Kutta methods, perhaps because of the larger storage requirements, as well as the complication that multistep methods are not self-starting and their difficulty in changing time step sizes during the computation, which are common disadvantages to all multistep methods, not just SSP methods. However, it is possible to design arbitrarily high order SSP multistep methods, while the order of SSP Runge–Kutta methods is bounded. Also, it was observed in [114, 115] that for applications requiring maximum principle satisfying methods or positivity preserving methods (see Chapter 11), SSP multistep methods maintain uniform high order accuracy better than SSP Runge–Kutta methods. This is likely the effect of the first order stage order of the Runge–Kutta methods. For these reasons, we expect that SSP multistep methods will be used more often in such applications in the future.

8.1 Bounds and barriers

8.1.1 *Explicit methods*

Explicit SSP multistep methods have no order barrier: this was proved in [89, 66]. However, for a fixed number of steps there is a limit to the attainable order.

Observation 8.1. For $s \geq 2$, there is no s step, order s SSP explicit linear multistep method with all non-negative β_i, and there is no s step SSP method (with positive SSP coefficient) of order $(s + 1)$ (even if negative coefficients are allowed).

Proof. Using the order conditions above, we see that an r order accurate method satisfies

$$\sum_{i=1}^{s} p(i)\alpha_i = \sum_{i=1}^{s} p'(i)\beta_i, \qquad (8.6)$$

for any polynomial $p(x)$ of degree at most r satisfying $p(0) = 0$.

When $r = s + 1$, we could choose

$$p(x) = \int_0^x q(t)dt, \qquad q(t) = \prod_{i=1}^{s}(i - t). \qquad (8.7)$$

Clearly $p'(i) = q(i) = 0$ for $i = 1, ..., s$. We also claim (and prove below) that all the values $p(i)$ for $i = 1, \ldots, s$, are positive. With this choice of p in (8.6), its right-hand side vanishes, while the left-hand side is strictly positive if all $\alpha_i \geq 0$, leading to a contradiction.

When $r = s$, we could choose

$$p(x) = x(s - x)^{s-1}.$$

Clearly $p(i) \geq 0$ for $i = 1, ..., s$, with the equality holding only for $i = s$. On the other hand, $p'(i) = s(1 - i)(s - i)^{s-2} \leq 0$, with the equality's holding only for $i = 1$ and $i = s$. Hence (8.6) would have a negative right side and a positive left side and would not be an equality, if all α_i and β_i are non-negative, unless the only nonzero entries are α_s, β_1 and β_s. In this special case we require $\alpha_s = 1$ and $\beta_1 = 0$ to get a positive SSP coefficient. The first two order conditions in (8.5) now leads to $\beta_s = s$ and $2\beta_s = s$, which cannot be simultaneously satisfied.

We conclude the proof by showing the inequality

$$p(i) = \int_0^i q(t)dt > 0, \qquad q(t) := \prod_{i=1}^s (i - t).$$

Observe that $q(t)$ oscillates between being positive on the even intervals $I_0 = (0,1), I_2 = (2,3), \ldots$ and being negative on the odd intervals, $I_1 = (1,2), I_3 = (3,4), \ldots.$ The positivity of the $p(i)$'s for $i \leq (s+1)/2$ follows from the fact that the integral of $q(t)$ over each pair of consecutive intervals is positive, at least for the first $[(s+1)/2]$ intervals,

$$\begin{aligned}
p(2k+2) - p(2k) &= \int_{I_{2k}} |q(t)|dt - \int_{I_{2k+1}} |q(t)|dt \\
&= \int_{I_{2k}} - \int_{I_{2k+1}} |(1-t)(2-t)\ldots(s-t)|dt \\
&= \int_{I_{2k}} |(1-t)(2-t)\ldots(s-1-t)| \times (|(s-t)| - |t|)dt \\
&> 0 \qquad 2k+1 \leq (s+1)/2.
\end{aligned}$$

For the remaining intervals, we note the symmetry of $q(t)$ w.r.t. the midpoint $(s+1)/2$, i.e. $q(t) = (-1)^s q(s + 1 - t)$, which enables us to write for $i > (s+1)/2$

$$\begin{aligned}
p(i) &= \int_0^{(s+1)/2} q(t)dt + (-1)^s \int_{(s+1)/2}^i q(s+1-t)dt \\
&= \int_0^{(s+1)/2} q(t)dt + (-1)^s \int_{s+1-i}^{(s+1)/2} q(t')dt'. \qquad (8.8)
\end{aligned}$$

Thus, if s is odd then $p(i) = p(s+1-i) > 0$ for $i > (s+1)/2$. If s is even, then the second integral on the right of (8.8) is positive for odd i's, since it starts with a positive integrand on the even interval, I_{s+1-i}. And finally, if s is even and i is odd, then the second integral starts with a negative contribution from its first integrand on the odd interval, I_{s+1-i}, while the remaining terms that follow cancel in pairs as before; a straightforward computation shows that this first negative contribution is compensated by the positive gain from the first pair, i.e.

$$p(s + 2 - i) > \int_0^2 q(t)dt + \int_{s+1-i}^{s+2-i} q(t)dt > 0, \qquad s \text{ even, } i \text{ odd.}$$

This concludes the proof.   □

The first part of Observation 8.1 is not very sharp for larger values of s, as can be seen from Tables 8.1 and 8.2. For instance, the minimum number of steps required for a tenth order SSP method is 22 – far more than the minimum of 12 indicated by the observation. On the other hand, results in Chapter 10 indicate that the second part of the observation is sharp.

We now turn our attention to the problem of finding explicit linear multistep methods with optimal SSP coefficients. We saw in Remark 5.2 that any explicit multistep method has threshold factor $R \leq 1$, which means that the SSP coefficient also satisfies $C \leq 1$. In fact, for explicit multistep methods it turns out that conditions for strong stability preservation are the same whether one considers linear or nonlinear problems, so that $C = R$ for this class of methods. The bound $C = 1$ is of course achieved by the forward Euler method. For higher order methods, we have the following bound on C, which was proved by Lenferink (Theorem 2.2 in [66]):

Observation 8.2. The SSP coefficient of an s step explicit linear multistep method of order $p > 1$, satisfies

$$C \leq \frac{s - p}{s - 1}. \tag{8.9}$$

Note that the first part of Observation 8.1 is a special case of this more general result.

8.1.2   *Implicit methods*

As before, we turn to implicit methods in the hope that we can obtain larger SSP coefficients. Once again, we find that the allowable SSP coefficient for implicit multistep methods of order $p > 1$ is only twice that for explicit methods. Whereas for Runge–Kutta methods this result is a conjecture, in the case of implicit multistep methods it was proved in [67, 48].

Observation 8.3. Any implicit multistep method of order $p > 1$ has SSP coefficient no greater than two.

It is interesting to note that this bound is actually obtained, for example, by the trapezoidal method. If we compare the efficiency of the trapezoidal method, which uses only one function evaluation and has SSP coefficient $\mathcal{C} = 2$, with that of the explicit Runge–Kutta method SSPRK (2,2), which requires two function evaluations and has SSP coefficient $\mathcal{C} = 1$, we notice that the explicit method requires four times as many function evaluations per unit time. However, the cost of solving the implicit system of equations is usually greater than the cost of four explicit function evaluations, so that the explicit method is more computationally efficient.

8.2 Explicit SSP multistep methods using few stages

In this section we provide a few examples of commonly used SSP multistep methods. Comparing low order SSP linear multistep methods with few steps and low order SSP Runge–Kutta methods with few stages, methods of both types have similar effective SSP coefficients, but Runge–Kutta methods with many stages have much better effective SSP coefficients than linear multistep methods with many steps; the Runge–Kutta methods also require less storage. However, between these two classes, only SSP linear multistep methods are available if an accuracy of greater than fourth order is required.

8.2.1 *Second order methods*

Explicit SSPMS(3,2): The three-step method in [91], given by

$$u^{n+1} = \frac{3}{4} \left(u^n + 2\Delta t F(u^n) \right) + \frac{1}{4} u^{n-2} \tag{8.10}$$

is second order accurate and has SSP coefficient $\mathcal{C} = \frac{1}{2}$.

This method is optimal, as the SSP coefficient satisfies the bound (8.9). To prove this directly, consider that the coefficients of the second order two-step method are given by:

$$\alpha_1 = \frac{1}{2} \left(6 - 3\beta_1 - \beta_2 + \beta_3 \right)$$
$$\alpha_2 = -3 + 2\beta_1 - 2\beta_3$$
$$\alpha_3 = \frac{1}{2} \left(2 - \beta_1 + \beta_2 + 3\beta_3 \right).$$

For $\mathcal{C} > \frac{1}{2}$ we need $\frac{\alpha_k}{\beta_k} > \frac{1}{2}$ $\quad \forall k$.

$$2\alpha_1 > \beta_1 \Rightarrow 6 - 4\beta_1 - \beta_2 + \beta_3 > 0$$
$$2\alpha_2 > \beta_2 \Rightarrow -6 + 4\beta_1 - \beta_2 - 4\beta_3 > 0.$$

This means that

$$\beta_2 - \beta_3 < 6 - 4\beta_1 < -\beta_2 - 4\beta_3 \Rightarrow 2\beta_2 < -3\beta_3.$$

Thus, we would have a negative β.

For multistep methods only one computation is required per time step, so that the effective SSP coefficient is $\mathcal{C}_{\text{eff}} = \mathcal{C} = \frac{1}{2}$. This method is apparently as efficient as the best two-stage, second order SSP Runge–Kutta method in Chapter 2. Of course, the storage requirement here is larger, as one must store the solution in the three previous time steps u^n, u^{n-1} and u^{n-2} in order to compute u^{n+1}.

In the case of Runge–Kutta methods, we increased the effective SSP coefficient by using more stages. Increasing the number of steps may improve the time step restriction for SSP multistep methods. Adding steps increases the SSP coefficient but does not require additional computation, only additional storage.

Explicit SSPMS(4,2): [91] The optimal four-step second order method is given by

$$u^{n+1} = \frac{8}{9}\left(u^n + \frac{3}{2}\Delta t F(u^n)\right) + \frac{1}{9}u^{n-3}. \tag{8.11}$$

This method has a SSP coefficient $\mathcal{C} = \frac{2}{3}$ and effective SSP coefficient $\mathcal{C}_{\text{eff}} = \frac{2}{3}$. Once again, the bound (8.9) allows us to conclude that this method is optimal.

Explicit SSPMS(s,2): The two methods above are the first members of a family of optimal s-step second order methods, which for $s > 2$ has coefficients [66]

$$\alpha_1 = \frac{(s-1)^2 - 1}{(s-1)^2}, \qquad \alpha_s = \frac{1}{(s-1)^2}, \qquad \beta_1 = \frac{s}{s-1}$$

and SSP coefficient $\mathcal{C} = \frac{s-2}{s-1}$. (As usual, the unlisted coefficients are zero.)

8.2.2　*Third order methods*

Explicit SSPMS(4,3): The optimal four-step third order method in [91] is given by

$$u^{n+1} = \frac{16}{27}\left(u^n + 3\Delta t F(u^n)\right) + \frac{11}{27}\left(u^{n-3} + \frac{12}{11}\Delta t F(u^{n-3})\right). \qquad (8.12)$$

It has SSP coefficient $\mathcal{C} = \frac{1}{3}$, which is clearly optimal as it matches the bound (8.9). To prove directly that this method is optimal, observe that the coefficients of the third order three-step method are given by:

$$\alpha_1 = \frac{1}{6}\left(24 - 11\beta_1 - 2\beta_2 + \beta_3 - 2\beta_4\right)$$

$$\alpha_2 = -6 + 3\beta_1 - \frac{1}{2}\beta_2 - \beta_3 + \frac{3}{2}\beta_4$$

$$\alpha_3 = 4 - \frac{3}{2}\beta_1 + \beta_2 + \frac{1}{2}\beta_3 - 3\beta_4$$

$$\alpha_4 = \frac{1}{6}\left(-6 + 2\beta_1 - \beta_2 + 2\beta_3 + 11\beta_4\right).$$

For $\mathcal{C} > \frac{1}{3}$ we need $\frac{\alpha_k}{\beta_k} > \frac{1}{3}$ $\forall k$. This implies:

$$72 - 39\beta_1 - 6\beta_2 + 3\beta_3 - 6\beta_4 > 0$$

$$-72 + 36\beta_1 - 10\beta_2 - 12\beta_3 + 18\beta_4 > 0$$

$$72 - 27\beta_1 + 18\beta_2 + 3\beta_3 - 54\beta_4 > 0$$

$$-72 + 24\beta_1 - 12\beta_2 + 24\beta_3 + 108\beta_4 > 0$$

combining these we get:

$$-\frac{46}{3}\beta_2 - 12\beta_3 > 0,$$

which implies a negative β.

Its effective SSP coefficient $\mathcal{C}_{\text{eff}} = \frac{1}{3}$ shows that this method is as efficient as the best three-stage, third order SSP Runge–Kutta method SSPRK(3,3) of Chapter 2.

Explicit SSPMS(5,3): The optimal five-step third order method is

$$u^{n+1} = \frac{25}{32}u^n + \frac{25}{16}\Delta t F(u^n) + \frac{7}{32}u^{n-4} + \frac{5}{16}\Delta t F(u^{n-4}). \qquad (8.13)$$

It has SSP coefficient $\mathcal{C} = \frac{1}{2}$, which matches the bound (8.9).

Explicit SSPMS(6,3)$_1$: The six-step method in [91] was given by

$$u^{n+1} = \frac{108}{125}\left(u^n + \frac{5}{3}\Delta t F(u^n)\right) + \frac{17}{125}\left(u^{n-5} + \frac{30}{17}\Delta t F(u^{n-5})\right) \qquad (8.14)$$

is third order accurate and SSP with a SSP coefficient $C = \frac{17}{30} \approx 0.57$. This is slightly less than the bound (8.9), which is $\frac{3}{5}$.

Explicit SSPMS(6,3)$_2$: A better method was found in [56, 57] using a modification of the optimization procedure in Chapter 4. Its coefficients are

$$\alpha_1 = 0.850708871672579 \quad \beta_1 = 1.459638436015276$$
$$\alpha_5 = 0.030664864534383 \quad \beta_5 = 0.052614491749200$$
$$\alpha_6 = 0.118626263793039 \quad \beta_6 = 0.203537849338252.$$

This method has SSP coefficient $C = 0.5828$, which is closer to the bound.

8.2.3 Fourth order methods

Explicit SSPMS(6,4): This method was found numerically in [56, 57].

$$\alpha_1 = 0.342460855717007 \quad \beta_1 = 2.078553105578060,$$
$$\alpha_4 = 0.191798259434736 \quad \beta_4 = 1.164112222279710$$
$$\alpha_5 = 0.093562124939008 \quad \beta_5 = 0.567871749748709$$
$$\alpha_6 = 0.372178759909247.$$

It has SSP coefficient $C = 0.1648$.

8.3 Optimal methods of higher order and more steps

8.3.1 Explicit methods

Optimal contractive explicit linear multistep methods were investigated by Lenferink [66], who discovered many interesting properties of these methods and computed optimal methods for up to 20 steps and seventh order accuracy. A general, fast algorithm for computing such methods was given in [56, 57]. Results for up to 50 steps are reproduced here in Tables 8.1 and 8.2. These are also optimal threshold factors.

8.3.2 Implicit methods

Lenferink determined the optimal SSP implicit linear multistep methods of up to 20 steps and order eight [67]. The algorithm of [56, 57] also applies here. SSP coefficients of optimal methods with up to 50 steps are reproduced in Tables 8.3 and 8.4.

Table 8.1   SSP coefficients (also the optimal threshold factors) of optimal explicit linear multistep methods up to order 7

$s \backslash p$	1	2	3	4	5	6	7
1	1.000						
2	1.000						
3	1.000	0.500					
4	1.000	0.667	0.333				
5	1.000	0.750	0.500	0.021			
6	1.000	0.800	0.583	0.165			
7	1.000	0.833	0.583	0.282	0.038		
8	1.000	0.857	0.583	0.359	0.145		
9	1.000	0.875	0.583	0.393	0.228		
10	1.000	0.889	0.583	0.421	0.282	0.052	
11	1.000	0.900	0.583	0.443	0.317	0.115	
12	1.000	0.909	0.583	0.460	0.345	0.175	0.018
13	1.000	0.917	0.583	0.474	0.370	0.210	0.077
14	1.000	0.923	0.583	0.484	0.390	0.236	0.116
15	1.000	0.929	0.583	0.493	0.406	0.259	0.154
16	1.000	0.933	0.583	0.501	0.411	0.276	0.177
17	1.000	0.938	0.583	0.507	0.411	0.291	0.198
18	1.000	0.941	0.583	0.513	0.411	0.304	0.217
19	1.000	0.944	0.583	0.517	0.411	0.314	0.232
20	1.000	0.947	0.583	0.521	0.411	0.322	0.246
21	1.000	0.950	0.583	0.525	0.411	0.330	0.259
22	1.000	0.952	0.583	0.528	0.411	0.337	0.269
23	1.000	0.955	0.583	0.531	0.411	0.342	0.278
24	1.000	0.957	0.583	0.534	0.411	0.347	0.286
25	1.000	0.958	0.583	0.536	0.411	0.351	0.294
26	1.000	0.960	0.583	0.538	0.411	0.354	0.301
27	1.000	0.962	0.583	0.540	0.411	0.358	0.307
28	1.000	0.963	0.583	0.542	0.411	0.360	0.312
29	1.000	0.964	0.583	0.543	0.411	0.363	0.317
30	1.000	0.966	0.583	0.545	0.411	0.365	0.319
31	1.000	0.967	0.583	0.546	0.411	0.368	0.319
32	1.000	0.968	0.583	0.548	0.411	0.370	0.319
33	1.000	0.969	0.583	0.549	0.411	0.371	0.319
34	1.000	0.970	0.583	0.550	0.411	0.373	0.319
35	1.000	0.971	0.583	0.551	0.411	0.375	0.319
36	1.000	0.971	0.583	0.552	0.411	0.376	0.319
37	1.000	0.972	0.583	0.553	0.411	0.377	0.319
38	1.000	0.973	0.583	0.554	0.411	0.378	0.319
39	1.000	0.974	0.583	0.555	0.411	0.379	0.319
40	1.000	0.974	0.583	0.555	0.411	0.380	0.319
41	1.000	0.975	0.583	0.556	0.411	0.381	0.319
42	1.000	0.976	0.583	0.557	0.411	0.382	0.319
43	1.000	0.976	0.583	0.557	0.411	0.383	0.319
44	1.000	0.977	0.583	0.558	0.411	0.384	0.319
45	1.000	0.977	0.583	0.559	0.411	0.385	0.319
46	1.000	0.978	0.583	0.559	0.411	0.385	0.319
47	1.000	0.978	0.583	0.560	0.411	0.386	0.319
48	1.000	0.979	0.583	0.560	0.411	0.387	0.319
49	1.000	0.979	0.583	0.561	0.411	0.387	0.319
50	1.000	0.980	0.583	0.561	0.411	0.388	0.319

Table 8.2   SSP coefficients (also the optimal threshold factors) of optimal explicit linear multistep methods of order 8-15

$s\backslash p$	8	9	10	11	12	13	14	15
15	0.012							
16	0.044							
17	0.075							
18	0.106	0.003						
19	0.128	0.035						
20	0.148	0.063						
21	0.163	0.082						
22	0.177	0.100	0.010					
23	0.190	0.116	0.031					
24	0.201	0.131	0.048					
25	0.210	0.145	0.063					
26	0.220	0.155	0.078	0.012				
27	0.228	0.165	0.091	0.027				
28	0.234	0.174	0.105	0.042				
29	0.240	0.184	0.116	0.055				
30	0.246	0.191	0.125	0.066	0.002			
31	0.250	0.199	0.134	0.079	0.014			
32	0.255	0.205	0.142	0.089	0.026			
33	0.259	0.211	0.150	0.097	0.036			
34	0.262	0.216	0.157	0.106	0.047			
35	0.266	0.221	0.164	0.114	0.057	0.005		
36	0.269	0.225	0.169	0.121	0.066	0.016		
37	0.271	0.229	0.175	0.128	0.074	0.027		
38	0.274	0.233	0.180	0.134	0.082	0.036		
39	0.276	0.237	0.185	0.140	0.089	0.044		
40	0.278	0.240	0.189	0.146	0.096	0.052	0.003	
41	0.280	0.243	0.193	0.152	0.103	0.059	0.011	
42	0.281	0.246	0.197	0.156	0.109	0.066	0.020	
43	0.283	0.249	0.200	0.161	0.114	0.073	0.027	
44	0.284	0.251	0.203	0.165	0.119	0.079	0.034	
45	0.286	0.253	0.206	0.169	0.124	0.085	0.040	0.000
46	0.287	0.256	0.209	0.172	0.129	0.090	0.047	0.008
47	0.288	0.258	0.211	0.176	0.133	0.095	0.053	0.015
48	0.289	0.260	0.213	0.179	0.138	0.100	0.058	0.022
49	0.290	0.261	0.215	0.182	0.141	0.105	0.064	0.028
50	0.291	0.261	0.218	0.185	0.145	0.109	0.069	0.034

8.4   Starting methods

The restrictive SSP coefficients observed in the SSP multistep methods are not surprising, considering that we require the SSP property to hold for *arbitrary* starting values. An illustration of the difficulty is given in [49]: Consider the simple example of the well-known BDF2 method applied to

Table 8.3   Optimal SSP coefficients for implicit linear multistep methods up to order 8.

$s \backslash p$	1	2	3	4	5	6	7	8
1	∞	2.000						
2	∞	2.000	1.000					
3	∞	2.000	1.500	1.000				
4	∞	2.000	1.667	1.243	0.667			
5	∞	2.000	1.750	1.243	0.796	0.500		
6	∞	2.000	1.800	1.243	0.929	0.660	0.300	
7	∞	2.000	1.833	1.243	1.006	0.784	0.468	0.197
8	∞	2.000	1.857	1.243	1.052	0.868	0.550	0.345
9	∞	2.000	1.875	1.243	1.084	0.905	0.642	0.443
10	∞	2.000	1.889	1.243	1.106	0.905	0.690	0.533
11	∞	2.000	1.900	1.243	1.123	0.905	0.733	0.580
12	∞	2.000	1.909	1.243	1.136	0.905	0.764	0.625
13	∞	2.000	1.917	1.243	1.147	0.905	0.781	0.662
14	∞	2.000	1.923	1.243	1.155	0.905	0.795	0.692
15	∞	2.000	1.929	1.243	1.162	0.905	0.806	0.714
16	∞	2.000	1.933	1.243	1.168	0.905	0.815	0.719
17	∞	2.000	1.938	1.243	1.174	0.905	0.823	0.719
18	∞	2.000	1.941	1.243	1.178	0.905	0.829	0.719
19	∞	2.000	1.944	1.243	1.182	0.905	0.835	0.719
20	∞	2.000	1.947	1.243	1.186	0.905	0.839	0.719
21	∞	2.000	1.950	1.243	1.189	0.905	0.844	0.719
22	∞	2.000	1.952	1.243	1.192	0.905	0.847	0.719
23	∞	2.000	1.955	1.243	1.194	0.905	0.851	0.719
24	∞	2.000	1.957	1.243	1.197	0.905	0.853	0.719
25	∞	2.000	1.958	1.243	1.199	0.905	0.856	0.719
26	∞	2.000	1.960	1.243	1.201	0.905	0.858	0.719
27	∞	2.000	1.962	1.243	1.202	0.905	0.861	0.719
28	∞	2.000	1.963	1.243	1.204	0.905	0.862	0.719
29	∞	2.000	1.964	1.243	1.205	0.905	0.864	0.719
30	∞	2.000	1.966	1.243	1.207	0.905	0.866	0.719
31	∞	2.000	1.967	1.243	1.208	0.905	0.867	0.719
32	∞	2.000	1.968	1.243	1.209	0.905	0.869	0.719
33	∞	2.000	1.969	1.243	1.210	0.905	0.870	0.719
34	∞	2.000	1.970	1.243	1.211	0.905	0.871	0.719
35	∞	2.000	1.971	1.243	1.212	0.905	0.872	0.719
36	∞	2.000	1.971	1.243	1.213	0.905	0.873	0.719
37	∞	2.000	1.972	1.243	1.214	0.905	0.874	0.719
38	∞	2.000	1.973	1.243	1.215	0.905	0.875	0.719
39	∞	2.000	1.974	1.243	1.216	0.905	0.876	0.719
40	∞	2.000	1.974	1.243	1.217	0.905	0.877	0.719
41	∞	2.000	1.975	1.243	1.217	0.905	0.878	0.719
42	∞	2.000	1.976	1.243	1.218	0.905	0.879	0.719
43	∞	2.000	1.976	1.243	1.218	0.905	0.879	0.719
44	∞	2.000	1.977	1.243	1.219	0.905	0.880	0.719
45	∞	2.000	1.977	1.243	1.220	0.905	0.881	0.719
46	∞	2.000	1.978	1.243	1.220	0.905	0.881	0.719
47	∞	2.000	1.978	1.243	1.221	0.905	0.882	0.719
48	∞	2.000	1.979	1.243	1.221	0.905	0.882	0.719
49	∞	2.000	1.979	1.243	1.222	0.905	0.883	0.719
50	∞	2.000	1.980	1.243	1.222	0.905	0.883	0.719

the problem $u'(t) = 0$:

$$u_2 = \frac{4}{3}u_1 - \frac{1}{3}u_0.$$

Clearly, this method is not SSP (α_2 is negative!). In other words, it is not always possible to obtain $||u_2|| \leq ||u_0||$ whenever $||u_1|| \leq ||u_0||$. However, it is also clear that the only relevant choice for this problem is $u_1 = u_0$, and in this case we do obtain (trivially) $||u_2|| \leq ||u_0||$.

Using this idea, Hundsdorfer, Ruuth, and Spiteri [49] examined the required step size for several multistep methods with particular starting procedures. Rather than satisfying a strict monotonicity property, these methods guarantee the boundedness property

$$||u^n|| \leq M||u^0||$$

where M is a constant depending on the starting procedures. Methods of this type were further considered in [85, 48], and methods of up to sixth order were given with reasonably large time step coefficients, for example a three-step, third order method with $\mathcal{C} = 0.537$ and a four-step, fourth order method with $\mathcal{C} = 0.458$. This creative approach to SSP multistep methods demonstrates that the SSP criteria may sometimes be relaxed or replaced by other conditions on the method.

Using this same idea for implicit methods was not as successful. In [49], the authors studied the case of implicit two-step methods with different starting procedures, to see if this approach provides a benefit similar to that seen in explicit multistep methods. Even with suitable starting procedures, the step size restrictions for the implicit multistep methods are hardly better than those of explicit methods. Furthermore, Hundsdorfer and Ruuth [48] showed that, in fact, methods of this type with order greater than one are subject to the same maximal SSP coefficient of two. Thus, implicit SSP multistep methods feature step size restrictions that may be too severe for these methods to be efficient in practice.

Table 8.4 Optimal SSP coefficients for implicit linear multistep methods of orders 9-15.

$s \backslash p$	9	10	11	12	13	14	15
8	0.006						
9	0.206	0.024					
10	0.295	0.127					
11	0.393	0.203					
12	0.444	0.295	0.093				
13	0.485	0.353	0.171	0.042			
14	0.526	0.402	0.238	0.103			
15	0.551	0.438	0.303	0.163			
16	0.578	0.468	0.342	0.221	0.076		
17	0.593	0.493	0.368	0.273	0.128	0.028	
18	0.609	0.518	0.397	0.299	0.173	0.072	
19	0.623	0.535	0.420	0.328	0.223	0.111	
20	0.631	0.550	0.443	0.356	0.253	0.159	0.034
21	0.639	0.564	0.459	0.374	0.279	0.191	0.081
22	0.646	0.577	0.475	0.394	0.302	0.221	0.120
23	0.651	0.587	0.487	0.410	0.320	0.242	0.148
24	0.656	0.595	0.497	0.426	0.336	0.264	0.178
25	0.661	0.596	0.506	0.439	0.353	0.283	0.201
26	0.665	0.596	0.514	0.450	0.368	0.300	0.221
27	0.668	0.596	0.520	0.459	0.383	0.314	0.241
28	0.671	0.596	0.527	0.468	0.392	0.327	0.257
29	0.674	0.596	0.532	0.476	0.402	0.341	0.270
30	0.676	0.596	0.536	0.482	0.413	0.352	0.283
31	0.678	0.596	0.540	0.489	0.420	0.362	0.294
32	0.679	0.596	0.543	0.495	0.426	0.372	0.305
33	0.681	0.596	0.547	0.500	0.433	0.380	0.316
34	0.682	0.596	0.549	0.505	0.438	0.388	0.327
35	0.684	0.596	0.552	0.508	0.443	0.394	0.335
36	0.685	0.596	0.554	0.508	0.448	0.400	0.343
37	0.686	0.596	0.556	0.508	0.451	0.406	0.349
38	0.687	0.596	0.557	0.508	0.455	0.411	0.356
39	0.688	0.596	0.559	0.508	0.458	0.416	0.363
40	0.689	0.596	0.560	0.508	0.461	0.420	0.368
41	0.690	0.596	0.562	0.508	0.463	0.424	0.373
42	0.691	0.596	0.563	0.508	0.465	0.428	0.377
43	0.692	0.596	0.564	0.508	0.467	0.431	0.381
44	0.692	0.596	0.565	0.508	0.469	0.434	0.385
45	0.693	0.596	0.566	0.508	0.471	0.437	0.389
46	0.694	0.596	0.567	0.508	0.473	0.439	0.391
47	0.694	0.596	0.568	0.508	0.474	0.441	0.395
48	0.695	0.596	0.568	0.508	0.475	0.444	0.397
49	0.695	0.580	0.569	0.508	0.476	0.443	0.399
50	0.696	0.573	0.570	0.508	0.477	0.443	0.402

Chapter 9

SSP Properties of Multistep Multi-Stage Methods

In the previous chapters we saw that explicit SSP Runge–Kutta methods suffer from an order barrier of four. To obtain higher order explicit SSP time discretizations, methods which include both multiple steps and multiple stages can be considered. Such methods are referred to as general linear methods. In this chapter we describe the SSP theory for general linear methods.

There has been significant recent activity in this field. In [15] Constantinescu and Sandu considered two- and three-step Runge–Kutta methods, and generated SSP methods of order up to four which have a certificate of global optimality. Huang [47] found methods of many stages and up to seventh order with good SSP coefficients. In this chapter, we focus on the SSP theory for general linear methods and describe recent advances in high order SSP two-step Runge–Kutta methods. This material originally appeared in [58], and the interested reader is referred to that work for further details.

9.1 SSP theory of general linear methods

In this section, we review the theory of strong stability preservation for general linear methods [96]. A general linear method can be written in the form

$$w_i^n = \sum_{j=1}^{l} s_{ij} x_j^n + \Delta t \sum_{j=1}^{m} t_{ij} F(w_j^n), \qquad (1 \le i \le m), \qquad (9.1a)$$

$$x_j^{n+1} = w_{J_j}^n, \qquad (1 \le j \le l). \qquad (9.1b)$$

The terms x_j^n are the l input values available from previous steps, and the w_j^n are the m intermediate stages used in computing the current step.

Equation (9.1b) indicates which of these values are used as inputs in the next step. The method is determined by its coefficients s_{ij} and t_{ij}, which can be written in an $m \times l$ matrix $\mathbf{S}$ and an $m \times m$ matrix $\mathbf{T}$, respectively. Without loss of generality (see [96]) we assume that

$$\mathbf{S}e = e. \tag{9.2}$$

This implies that every stage is a consistent approximation to the solution at some time.

Runge–Kutta methods and multistep methods are subclasses of general linear methods, and can therefore be written in the form (9.1).

Example 9.1. An s-stage Runge–Kutta method with Butcher tableau $\mathbf{A}$ and $\boldsymbol{b}$ can be written in form (9.1) by taking $l = 1, m = s + 1, J = \{m\}$, and

$$\mathbf{S} = (1, 1, \dots, 1)^{\mathrm{T}}, \qquad \mathbf{T} = \begin{pmatrix} \mathbf{A} & 0 \\ \boldsymbol{b}^{\mathrm{T}} & 0 \end{pmatrix}. \tag{9.3}$$
$\square$

Example 9.2. Linear multistep methods

$$u^{n+1} = \sum_{j=1}^{l} \alpha_j u^{n+1-j} + \Delta t \sum_{j=0}^{l} \beta_j F\left(u^{n+1-j}\right)$$

admit the Spijker form

$$\mathbf{S} = \begin{pmatrix} 1 & 0 & \dots & 0 \\ 0 & \ddots & \ddots & 0 \\ 0 & \ddots & \ddots & 0 \\ 0 & \ddots & \ddots & 1 \\ \alpha_1 & \alpha_2 & \dots & \alpha_l \end{pmatrix}, \quad \mathbf{T}^{(l+1)\times(l+1)} = \begin{pmatrix} 0 & 0 & \dots & 0 \\ 0 & \ddots & \ddots & 0 \\ 0 & \ddots & \ddots & 0 \\ 0 & \ddots & 0 & 0 \\ \beta_0 & \beta_1 & \dots & \beta_l \end{pmatrix}$$

where $\mathbf{S}$ is an $(l+1) \times l$ and $\mathbf{T}$ is an $(l+1) \times (l+1)$ matrix, and $J = \{2, \dots, l+1\}$.

$\square$

In order to analyze the SSP property of a general linear method (9.1), we proceed in a similar manner to the approach of Chapter 3. We first define the vector $\boldsymbol{f} = [F(w_1), F(w_2), \dots, F(w_m)]^{\mathrm{T}}$, so that (9.1a) can be written compactly as

$$\mathbf{w} = \mathbf{S}\mathbf{x} + \Delta t \mathbf{T}\boldsymbol{f}. \tag{9.4}$$

Adding $r\mathbf{T}\mathbf{y}$ to both sides of (9.4) gives

$$(\mathbf{I} + r\mathbf{T})\,\mathbf{w} = \mathbf{S}\mathbf{x} + r\mathbf{T}\left(\mathbf{w} + \frac{\Delta t}{r}f\right). \tag{9.5}$$

Assuming that the matrix on the left is invertible, we obtain

$$\mathbf{w} = (\mathbf{I} + r\mathbf{T})^{-1}\mathbf{S}\mathbf{x} + r(\mathbf{I} + r\mathbf{T})^{-1}\mathbf{T}\left(\mathbf{w} + \frac{\Delta t}{r}f\right)$$

$$= \mathbf{R}\mathbf{x} + \mathbf{P}\left(\mathbf{w} + \frac{\Delta t}{r}f\right), \tag{9.6}$$

where we have defined

$$\mathbf{P} = r(\mathbf{I} + r\mathbf{T})^{-1}\mathbf{T}, \quad \mathbf{R} = (\mathbf{I} + r\mathbf{T})^{-1}\mathbf{S} = (\mathbf{I} - \mathbf{P})\mathbf{S}. \tag{9.7}$$

Observe that, by the consistency condition (9.2), the row sums of $[\mathbf{R}\ \mathbf{P}]$ are each equal to one:

$$\mathbf{R}e + \mathbf{P}e = (\mathbf{I} - \mathbf{P})\mathbf{S}e + \mathbf{P}e = e - \mathbf{P}e + \mathbf{P}e = e. \tag{9.8}$$

Thus, if $\mathbf{R}$ and $\mathbf{P}$ have no negative entries, each stage w_i is given by a convex combination of the inputs x_j and the quantities $w_j + (\Delta t/r)F(w_j)$. In other words, this method is a convex combination of forward Euler steps. Hence any strong stability property of the forward Euler method is preserved by the method (9.4) under the time step restriction given by $\Delta t \le \mathcal{C}(\mathbf{S}, \mathbf{T})\Delta t_{\mathrm{FE}}$ where $\mathcal{C}(\mathbf{S}, \mathbf{T})$ is defined as

$$\mathcal{C}(\mathbf{S}, \mathbf{T}) = \sup_r \left\{r : (I + r\mathbf{T})^{-1} \text{ exists and } \boldsymbol{\alpha} \ge 0, \mathbf{R} \ge 0\right\},$$

where $\mathbf{P}$ and $\mathbf{R}$ are defined in (9.7). By the foregoing observation, it is clear that the SSP coefficient of method (9.6) is greater than or equal to $\mathcal{C}(\mathbf{S}, \mathbf{T})$.

To state precisely the conditions under which the SSP coefficient is, in fact, equal to $\mathcal{C}(\mathbf{S}, \mathbf{T})$, we must introduce the concept of reducibility. A Runge–Kutta method is said to be *reducible* if there exists a method with fewer stages that always produces the same output. A particularly simple kind of reducibility is known as *HS-reducibility*; a Runge–Kutta method is HS-reducible if two of its stages are identically equal. For general linear methods, the definition of reducibility employed in [96] is essentially an extension of HS-reducibility, but it is inconvenient for our present purposes. Instead we introduce the following definition of HS-reducibility for general linear methods.

Definition 9.1 (HS-Reducibility). *A method in form* (9.1) *is HS-reducible if there exist indices i, j such that all of the following hold:*

1. *Rows i and j of $\mathbf{S}$ are equal.*
2. *Rows i and j of $\mathbf{T}$ are equal.*
3. *Column i of $\mathbf{T}$ is not identically zero, or $\mathbf{w}_i$ is an element of $\mathbf{x}^n$.*
4. *Column j of $\mathbf{T}$ is not identically zero, or $\mathbf{w}_j$ is an element of $\mathbf{x}^n$.*

Otherwise, we say the method is HS-irreducible.

The definition of reducibility used in Theorem 2.7 of [96] does not include the third and fourth conditions from Definition 9.1. However, Definition 9.1 is a more natural extension of the HS-reducibility concept from Runge–Kutta methods. For instance, the backward Euler method written in the usual form is HS-reducible by the definition from [96], but not by Definition 9.1.

A careful reading of the proof of Theorem 2.7 in [96] reveals that it still holds under Definition 9.1. Thus the following theorem is a slightly stronger statement.

Theorem 9.1. *(cf. Theorem 2.7 [96]) Let $\mathbf{S}, \mathbf{T}$ be an HS-irreducible representation of a general linear method. Then the SSP coefficient of the method is $\mathcal{C} = \mathcal{C}(\mathbf{S}, \mathbf{T})$.*

$\square$

9.2 Two-step Runge–Kutta methods

In the remainder of this chapter, we focus on a general class of two-step Runge–Kutta (TSRK) methods studied in [50, 7, 33, 109]. TSRK methods are a generalization of Runge–Kutta methods that include values and stages from the previous step:

$$y_i^n = d_i u^{n-1} + (1 - d_i)u^n + \Delta t \sum_{j=1}^{s} \hat{a}_{ij} F(y_j^{n-1}) + \Delta t \sum_{j=1}^{s} a_{ij} F(y_j^n), \quad 1 \leq i \leq s$$
$$\tag{9.9a}$$

$$u^{n+1} = \theta u^{n-1} + (1 - \theta)u^n + \Delta t \sum_{j=1}^{s} \hat{b}_j F(y_j^{n-1}) + \Delta t \sum_{j=1}^{s} b_j F(y_j^n). \tag{9.9b}$$

Here u^n and u^{n-1} denote solution values at the times $t = n\Delta t$ and $t = (n-1)\Delta t$, while the values y_i^n are intermediate stages used to compute the solution at the next time step.

General TSRK methods (9.9) can be written in Spijker form as follows: set $m = 2s + 2$, $l = s + 2$, $J = \{s + 1, s + 2, \ldots, 2s + 2\}$, and

$$\mathbf{x}^n = \left(u^{n-1}, y_1^{n-1}, \ldots, y_s^{n-1}, u^n\right)^{\mathrm{T}}, \tag{9.10a}$$

$$\mathbf{w}^n = \left(y_1^{n-1}, y_2^{n-1}, \ldots, y_s^{n-1}, u^n, y_1^n, y_2^n, \ldots, y_s^n, u^{n+1}\right)^{\mathrm{T}}, \tag{9.10b}$$

$$\mathbf{S} = \begin{pmatrix} 0 & \mathbf{I} & 0 \\ 0 & 0 & 1 \\ d & 0 & e - d \\ \theta & 0 & 1 - \theta \end{pmatrix}, \qquad \mathbf{T} = \begin{pmatrix} 0 & 0 & 0 & 0 \\ 0 & 0 & 0 & 0 \\ \hat{\mathbf{A}} & 0 & \mathbf{A} & 0 \\ \hat{b} & 0 & b^{\mathrm{T}} & 0 \end{pmatrix}. \tag{9.10c}$$

Here we are interested in TSRK methods that have the strong stability preserving property. As we will prove in Theorem 9.3, this greatly reduces the set of methods relevant to our study. Except in special cases, the method (9.9) cannot be strong stability preserving unless all of the coefficients $\hat{a}_{ij}, \hat{b}_j$ are identically zero. A brief explanation of this requirement is as follows. Since method (9.9) does not include terms of the form y_i^{n-1}, it is not possible to write a stage of method (9.9) as a convex combination of forward Euler steps if the stage includes terms of the form $F(y_i^{n-1})$. This is because those stages depend on u^{n-2}, which is not available in a two-step method.

Hence we are led to consider simpler methods of the following form (compare p. 362 of [32]). We call these **Type I** methods:

$$y_i^n = d_i u^{n-1} + (1 - d_i)u^n + \Delta t \sum_{j=1}^{s} a_{ij} F(y_j^n), \qquad 1 \le i \le s, \tag{9.11a}$$

$$u^{n+1} = \theta u^{n-1} + (1 - \theta)u^n + \Delta t \sum_{j=1}^{s} b_j F(y_j^n). \tag{9.11b}$$

Now consider the special case in which the method (9.9) has some stage y_i^n identically equal to u^n. Then including terms proportional to $F(u^n)$ will not prevent the method from being written as a convex combination of forward Euler steps; furthermore, since $y_i^{n-1} = u^{n-1}$, terms of the form $F(u^{n-1})$ can also be included. This leads to what we will call **Type II** methods, which have the form:

$$y_i^n = d_i u^{n-1} + (1 - d_i)u^n + \hat{a}_{i1}\Delta t F(u^{n-1}) + \Delta t \sum_{j=1}^{s} a_{ij} F(y_j^n), \quad 1 \le i \le s, \tag{9.12a}$$

$$u^{n+1} = \theta u^{n-1} + (1 - \theta)u^n + \hat{b}_1 \Delta t F(u^{n-1}) + \Delta t \sum_{j=1}^{s} b_j F(y_j^n). \tag{9.12b}$$

Example 9.3. Type I methods (9.11) can be written in Spijker form with $m = s + 2$, $l = 2$, $J = \{1, s + 2\}$, and

$$\mathbf{x}^n = \left(u^{n-1}, u^n\right)^{\mathrm{T}}, \quad \mathbf{w}^n = \left(u^n, y_1^n, y_2^n, \ldots, y_s^n, u^{n+1}\right)^{\mathrm{T}} \tag{9.13a}$$

$$\mathbf{S} = \begin{pmatrix} 0 & 1 \\ d & e - d \\ \theta & 1 - \theta \end{pmatrix}, \quad \mathbf{T} = \begin{pmatrix} 0 & \mathbf{0} & 0 \\ \mathbf{0} & \mathbf{A} & \mathbf{0} \\ 0 & \mathbf{b}^{\mathrm{T}} & 0 \end{pmatrix}. \tag{9.13b}$$

□

Example 9.4. Type II methods (9.12) can be written in the simple Spijker representation (with $m = s + 2$, $l = 2$, and $J = \{2, s + 2\}$):

$$\mathbf{x}^n = \left(u^{n-1}, u^n\right)^{\mathrm{T}}, \quad \mathbf{w}^n = \left(u^{n-1}, y_1^n, y_2^n, \ldots, y_s^n, u^{n+1}\right)^{\mathrm{T}} \tag{9.14a}$$

$$\mathbf{S} = \begin{pmatrix} 1 & 0 \\ d & e - d \\ \theta & 1 - \theta \end{pmatrix}, \quad \mathbf{T} = \begin{pmatrix} 0 & \mathbf{0} & 0 \\ \hat{a}_1 & \mathbf{A} & \mathbf{0} \\ \hat{b}_1 & \mathbf{b}^{\mathrm{T}} & 0 \end{pmatrix}. \tag{9.14b}$$

□

9.2.1 *Conditions and barriers for SSP two-step Runge–Kutta methods*

In light of Theorem 9.1, we are interested in methods with $\mathcal{C}(\mathbf{S}, \mathbf{T}) > 0$. The following lemma characterizes such methods.

Lemma 9.1. *(see Theorem 2.2(i)] in [96]) $\mathcal{C}(\mathbf{S}, \mathbf{T}) > 0$ if and only if all of the following hold:*

$$\mathbf{S} \geq 0, \tag{9.15a}$$

$$\mathbf{T} \geq 0, \tag{9.15b}$$

$$\mathrm{Inc}(\mathbf{TS}) \leq \mathrm{Inc}(\mathbf{S}), \tag{9.15c}$$

$$\mathrm{Inc}(\mathbf{T}^2) \leq \mathrm{Inc}(\mathbf{T}), \tag{9.15d}$$

where all the inequalities are element-wise and the incidence matrix of a matrix $\mathbf{M}$ with entries m_{ij} is

$$\mathrm{Inc}(\mathbf{M})_{ij} = \begin{cases} 1 & if\ m_{ij} \neq 0 \\ 0 & if\ m_{ij} = 0. \end{cases}$$

□

Combining Theorem 9.1 and Lemma 9.15, leads us to Type I and Type II methods.

Corollary 9.1. *Let $\mathbf{S}, \mathbf{T}$ be the coefficients of an s-stage two-step Runge–Kutta method (9.9) in the form (9.10), and suppose the method is HS-irreducible in this form and has positive SSP coefficient. Then the method can be written as an s-stage Type I method (9.11).*

Proof. This follows immediately from Theorem 9.1 and condition (9.15c) of Lemma 9.15, since, by the latter, $\mathcal{C}(\mathbf{S}, \mathbf{T}) > 0$ implies $\hat{\mathbf{A}} = \hat{\boldsymbol{b}} = \mathbf{0}$. Under this restriction, methods of the form (9.9) simplify to Type I methods (9.11).

$\square$

To apply Corollary 9.1, it is necessary to write a two-step Runge–Kutta method in HS-irreducible form. The next corollary deals with methods that are HS-reducible in the form (9.9).

Corollary 9.2. *Let* $\mathbf{S}, \mathbf{T}$ *be the coefficients of an s-stage two-step Runge–Kutta method* (9.9) *in the form* (9.10), *and suppose the method has positive SSP coefficient and that one of the stages y_j is identically equal to u^n. Then the method can be written as an s-stage Type II method* (9.12).

Proof. If necessary, reorder the stages so that $y_1^n = u^n$. Then rows $s + 1$ and $s + 2$ of $[\mathbf{S}\ \mathbf{T}]$ in the representation (9.10) are equal, so the method is HS-reducible. In this case, the method can be written in the irreducible Spijker form (noting that also $y_1^{n-1} = u^{n-1}$) as follows. Set $m = 2s + 1$, $l = s + 1$, and $J = \{s, s + 1, s + 2, \ldots, 2s + 2\}$, and

$$\mathbf{x}^n = \left(u^{n-1}, y_2^{n-1}, \ldots, y_s^{n-1}, u^n\right)^{\mathrm{T}} \tag{9.16a}$$

$$\mathbf{w}^n = \left(u^{n-1}, y_2^{n-1}, y_3^{n-1}, \ldots, y_s^{n-1}, y_1^n, y_2^n, \ldots, y_s^n, u^{n+1}\right)^{\mathrm{T}} \tag{9.16b}$$

$$\mathbf{S} = \begin{pmatrix} 1 & \mathbf{0} & 0 \\ \mathbf{0} & \mathbf{I} & \mathbf{0} \\ d & \mathbf{0} & e-d \\ \theta & \mathbf{0} & 1-\theta \end{pmatrix}, \qquad \mathbf{T} = \begin{pmatrix} 0 & \mathbf{0} & 0 & 0 \\ \mathbf{0} & \mathbf{0} & \mathbf{0} & \mathbf{0} \\ \hat{a}_1 & \hat{\mathbf{A}}_{2:s} & \mathbf{A} & \mathbf{0} \\ \hat{b}_1 & \hat{\boldsymbol{b}}_{2:s}^{\mathrm{T}} & \boldsymbol{b}^{\mathrm{T}} & 0 \end{pmatrix}. \tag{9.16c}$$

Here $\hat{a}_1$ and $\hat{b}_1$ represent the first column and first element of $\hat{\mathbf{A}}$ and $\hat{\boldsymbol{b}}$, respectively, while $\hat{\mathbf{A}}_{2:s}$ and $\hat{\boldsymbol{b}}_{2:s}^{\mathrm{T}}$ represent the remaining columns and remaining entries. Applying condition (9.15c) of Lemma 9.15 to the representation (9.16), we find that $\hat{\mathbf{A}}_{2:s}$ and $\hat{\boldsymbol{b}}_{2:s}^{\mathrm{T}}$ must vanish, but $\hat{a}_1$ and δ_1 may be non-zero. The resulting methods are Type II methods (9.12). $\square$

Theorem 9.2. *All two-step Runge–Kutta methods with positive SSP coefficient are of Type I* (9.11) *or Type II* (9.12).

Proof. If the method is HS-irreducible in form (9.10), apply Corollary 9.1. If the method is reducible in this form, then either $y_i^n = y_j^n$ for some i, j, or $y_j^n = u^n$, for some j (or both). In case $y_i^n = y_j^n$, one can simply remove stage y_i and adjust $\mathbf{A}, \hat{\mathbf{A}}$ correspondingly; this can be repeated as necessary until the method is irreducible or the reducibility is due only to the case $y_j^n = u^n$. Then apply Corollary 9.2.

$\square$

It is known that irreducible strong stability preserving Runge–Kutta methods have positive stage coefficients, $a_{ij} \geq 0$ and strictly positive weights, $b_j > 0$. The following theorem shows that similar properties hold for SSP two-step Runge–Kutta methods methods. The theorem and its proof are very similar to Theorem 4.2 in [62] (see also Theorem 2.2(i) in [96]).

In the proof, we will also use the concept of irreducibility introduced in Section 3.2.2. Recall that a method is said to be *DJ-reducible* if it involves one or more stages whose value does not affect the output. If a method is neither HS-reducible nor DJ-reducible, we say it is *irreducible*.

Theorem 9.3. *The coefficients of an irreducible two-step Runge–Kutta method with positive SSP coefficient satisfy the following bounds:*
(i) $\mathbf{A} \geq 0, \boldsymbol{b} \geq 0, 0 \leq \boldsymbol{d} \leq 1$, *and* $0 \leq \theta \leq 1$.
(ii) $\boldsymbol{b} > 0$.
All of these inequalities should be interpreted componentwise.

Proof.   By Theorem 9.2, any such method can be written in either form (9.11) or form (9.12). According to Theorem 9.1, we just need to show that (i)-(ii) are necessary in order to have $\mathcal{C}(\mathbf{S}, \mathbf{T}) > 0$. Condition (i) follows from conditions (9.15a) and (9.15b).

Furthermore, condition (9.15d) of Lemma 9.15 means that if $b_j = 0$ for some j then

$$\sum_i b_i a_{ij} = 0. \qquad (9.17)$$

Since $\mathbf{A}, \boldsymbol{b}$ are non-negative, (9.17) implies that either b_i or a_{ij} is zero for each value of i.

Now partition the set $\mathcal{S} = \{1, 2, \ldots, s\}$ into $\mathcal{S}_1, \mathcal{S}_2$ such that $b_j > 0$ for all $j \in \mathcal{S}_1$ and $b_j = 0$ for all $j \in \mathcal{S}_2$. Then $a_{ij} = 0$ for all $i \in \mathcal{S}_1$ and $j \in \mathcal{S}_2$. This implies that the method is DJ-reducible, unless $\mathcal{S}_2 = \emptyset$.

$\square$

Just as for Runge–Kutta methods, we find that SSP TSRK methods have a lower bound on the stage order, and an upper bound on the over-all order. We state these observations here without proof, and refer the interested reader to [58].

Observation 9.1. Any irreducible two-step Runge–Kutta method (9.9) of order p with positive SSP coefficient has stage order at least $\lfloor \frac{p-1}{2} \rfloor$.

Observation 9.2. The order of an explicit SSP two-step Runge–Kutta method is at most eight. Furthermore, if the method has order greater than six $(p > 6)$, it must be of Type II.

Observation 9.3. Any (implicit) SSP two-step Runge–Kutta method with $p > 8$ must be of Type II.

9.3   Optimal two-step Runge–Kutta methods

In this section we present the optimization problem and the resulting optimal SSP two-step Runge–Kutta methods. The methods presented were found in [58] via numerical search using MATLAB's Optimization and Global Optimization toolboxes. Although the authors searched over Type I and Type II methods; the optimal methods found are of Type II in every case. While it is difficult to state with certainty that these methods are globally optimal, this search recovered the global optimum in every case for which it was already known.

9.3.1   *Formulating the optimization problem*

The optimization problem is formulated using Spijker's theory:

$$\max_{\mathbf{S},\mathbf{T}} r, \tag{9.18a}$$

$$\text{subject to} \quad \begin{cases} (\mathbf{I} + r\mathbf{T})^{-1}\mathbf{S} \geq 0, \\ (\mathbf{I} + r\mathbf{T})^{-1}\mathbf{T} \geq 0, \\ \Phi_p(\mathbf{S}, \mathbf{T}) = 0, \end{cases} \tag{9.18b}$$

where the inequalities are understood component-wise and $\Phi_p(\mathbf{S}, \mathbf{T})$ represents the order conditions up to order p. This formulation, implemented in MATLAB using a sequential quadratic programming approach (fmincon in the optimization toolbox), was used to find the methods given below.

In comparing methods with different numbers of stages, one is usually interested in the relative time advancement per computational cost. For this purpose, we define the *effective SSP coefficient*

$$\mathcal{C}_{\text{eff}}(\mathbf{S}, \mathbf{T}) = \frac{\mathcal{C}(\mathbf{S}, \mathbf{T})}{s}.$$

This normalization enables us to compare the cost of integration up to a given time using two-step schemes of order $p > 1$.

It should be remembered that the optimal SSP methods that we find in the classes of Type I and Type II two-step Runge–Kutta methods are in fact optimal over the larger class of methods (9.9). Also, because they do not use intermediate stages from previous time steps, special conditions on the starting method (important for methods of the form (9.9) [33, 109, 108]) are unnecessary. Instead, the method can be started with any Runge–Kutta method of the appropriate order.

9.3.2 *Efficient implementation of Type II SSP TSRKs*

The form (9.6), with $r = \mathcal{C}(\mathbf{S}, \mathbf{T})$, typically yields very sparse coefficient matrices for optimal Type II SSP TSRK methods. This form is useful for an efficient (in terms of storage) implementation. To facilitate the presentation of this form, we introduce a compact, unified notation for both types of methods. First we rewrite an s-stage Type II method (9.12) by including u^{n-1} as one of the stages:

$$y_0^n = u^{n-1},$$

$$y_1^n = u^n,$$

$$y_i^n = d_i u^{n-1} + (1 - d_i)u^n + \Delta t \sum_{j=0}^{s} a_{ij} F(y_j^n), \qquad 2 \le i \le s,$$

$$u^{n+1} = \theta u^{n-1} + (1 - \theta)u^n + \Delta t \sum_{j=0}^{s} b_j F(y_j^n).$$

Then both Type I and Type II methods can be written in the compact form

$$\boldsymbol{y}^n = \bar{\boldsymbol{d}} u^{n-1} + (\mathbf{e} - \bar{\boldsymbol{d}})u^n + \Delta t \bar{\mathbf{A}} \mathbf{f}^n, \tag{9.19a}$$

$$u^{n+1} = \theta u^{n-1} + (1 - \theta)u^n + \Delta t \bar{\boldsymbol{b}}^{\mathrm{T}} \mathbf{f}^n, \tag{9.19b}$$

where, for Type I methods

$$\boldsymbol{y}^n = [y_1^n, \dots, y_s^n]^{\mathrm{T}}, \quad \mathbf{f}^n = [F(y_1^n), \dots, F(y_s^n)]^{\mathrm{T}},$$

$$\bar{\boldsymbol{d}} = [d_1, d_2, \dots, d_s]^{\mathrm{T}}, \quad \bar{\boldsymbol{b}} = \boldsymbol{b}, \quad \bar{\mathbf{A}} = \mathbf{A};$$

and for Type II methods

$$\boldsymbol{y}^n = [u^{n-1}, u^n, y_2^n, \dots, y_s^n]^{\mathrm{T}}, \quad \mathbf{f}^n = [F(u^{n-1}), F(u^n), F(y_2^n), \dots, F(y_s^n)]^{\mathrm{T}},$$

$$\bar{\boldsymbol{d}} = [1, 0, d_2, \dots, d_s]^{\mathrm{T}}, \quad \bar{\boldsymbol{b}} = [\hat{b}_1 \ \boldsymbol{b}^{\mathrm{T}}]^{\mathrm{T}}, \quad \bar{\mathbf{A}} = \begin{pmatrix} 0 & \mathbf{0} \\ \hat{a}_1 & \mathbf{A} \end{pmatrix}.$$

The efficient form for implementation is:

$$y_i^n = \tilde{d}_i u^{n-1} + \left(1 - \tilde{d}_i - \sum_{j=0}^{s} q_{ij}\right) u^n + \sum_{j=0}^{s} q_{ij}\left(y_j^n + \frac{\Delta t}{r}F(y_j^n)\right),$$

$$\text{for } (1 \leq i \leq s),$$

$$(9.20\text{a})$$

$$u^{n+1} = \tilde{\theta} u^{n-1} + \left(1 - \tilde{\theta} - \sum_{j=0}^{s} \eta_j\right) u^n + \sum_{j=0}^{s} \eta_j\left(y_j^n + \frac{\Delta t}{r}F(y_j^n)\right),$$

$$(9.20\text{b})$$

where the coefficients are given by (using the relations (9.7)):

$$\mathbf{Q} = r\bar{\mathbf{A}}(\mathbf{I} + r\bar{\mathbf{A}})^{-1}, \qquad \boldsymbol{\eta} = r\mathbf{b}^{\mathrm{T}}(\mathbf{I} + r\bar{\mathbf{A}})^{-1},$$

$$\tilde{d} = \bar{d} - \mathbf{Q}\bar{d}, \qquad \tilde{\theta} = \theta - \boldsymbol{\eta}^{\mathrm{T}}\bar{d}.$$

When implemented in this form, many of the methods presented in the next section have modest storage requirements, despite using large numbers of stages. The analogous form for Runge–Kutta methods was used in [55].

In the following sections, we give the coefficients in the efficient form (9.20) for the numerically optimal methods.

9.3.3   Optimal methods of orders one to four

In the case of first order methods, one can do no better (in terms of effective SSP coefficient) than the forward Euler method. For orders two to four, SSP coefficients of optimal methods are listed in Table 9.1. We list these mainly for completeness, since SSP Runge–Kutta methods with good properties exist up to order four, and these are typically the methods one would choose.

In [56], upper bounds for the values in Table 9.1 are found by finding optimally contractive general linear methods for linear systems of ODEs. Comparing the present results to the two-step results from that work, we see that this upper bound is achieved (as expected) for all first and second order methods, and even for the two- and three-stage third order methods.

Two-step general linear methods of up to fourth order using up to four stages were found in [15], and again in [58]. In [15] these methods were found by software that guarantees global optimality.

The optimal s-stage, second order SSP two-step Runge–Kutta method is in fact both a Type I and Type II method, and was found in numerical

Table 9.1 Effective SSP coefficients $\mathcal{C}_{\text{eff}}$ of optimal explicit two-step Runge–Kutta methods of order two to four. Results known to be optimal are shown in **bold**

s \ p	2	3	4
2	**0.707**	**0.366**	
3	**0.816**	**0.550**	**0.286**
4	**0.866**	**0.578**	**0.399**
5	**0.894**	0.598	0.472
6	**0.913**	0.630	0.509
7	**0.926**	0.641	0.534
8	**0.935**	0.653	0.562
9	**0.943**	0.667	0.586
10	**0.949**	0.683	0.610

searches over methods of both forms. It has SSP coefficient $\mathcal{C} = \sqrt{s(s-1)}$ and nonzero coefficients

$$q_{i,i-1} = 1, \qquad\qquad (2 \leq i \leq s),$$

$$q_{s+1,s} = 2(\mathcal{C} - s + 1),$$

$$\tilde{d} = 0,$$

$$\tilde{\theta} = 2(s - \mathcal{C}) - 1.$$

Note that these methods have $\mathcal{C}_{\text{eff}} = \sqrt{\frac{s-1}{s}}$, whereas the corresponding optimal Runge–Kutta methods have $\mathcal{C}_{\text{eff}} = \frac{s-1}{s}$. For $s = 2, 3, 4$, this is a significant improvement in $\mathcal{C}_{\text{eff}}$: 41%, 22%, 15% larger, respectively, than the optimal explicit Runge–Kutta method with the same number of stages. Using the low-storage assumption introduced in [55], these methods can be implemented with just three storage registers, just one register more than is required for the optimal second order SSP Runge–Kutta methods.

The optimal nine-stage, third order method is remarkable in that it is a Runge–Kutta method. In other words, allowing the freedom of using an additional step does not improve the SSP coefficient in this case.

9.3.4 *Optimal methods of orders five to eight*

Table 9.2 lists effective SSP coefficients of numerically optimal two-step Runge–Kutta methods of orders five to eight. Although these methods require many stages, it should be remembered that high order Rung–Kutta methods also require many stages. Indeed, some of our SSP two-step Runge–Kutta methods methods have fewer stages than the minimum number required to achieve the corresponding order for a Runge–Kutta method (regardless of SSP considerations).

The fifth order methods present an unusual phenomenon: when the number of stages is allowed to be greater than eight, it is not possible to achieve a larger effective SSP coefficient than the optimal 8-stage method, even allowing as many as 12 stages. This appears to be accurate, and not simply due to failure of the numerical optimizer, since in the nine-stage case the optimization scheme recovers the apparently optimal method in less than one minute, but fails to find a better result after several hours.

The only existing SSP methods of order greater than four are the hybrid methods of Huang [47]. Comparing the best two-step Runge–Kutta methods of each order with the best hybrid methods of each order, one sees that the two-step Runge–Kutta methods have substantially larger effective SSP coefficients, as well as modest storage requirements.

The effective SSP coefficient is a fair metric for comparison between methods of the same order of accuracy. Furthermore, our 12-stage two-step Runge–Kutta methods have sparse coefficient matrices and can be implemented in the form (9.20) in a very efficient manner with respect to storage. Specifically, the fifth through eighth order methods of 12 stages require only 5, 7, 7, and 10 memory locations per unknown, respectively, under the low-storage assumption employed in [55, 59]. Typically the methods with fewer stages require the same or more storage, so there is no reason to prefer methods with fewer stages if they have lower effective SSP coefficients. Thus, for sixth through eighth order, the 12-stage methods seem preferable. We remark that these SSP methods even require less storage than what (non-SSP one step) Runge–Kutta methods of the corresponding order would typically use.

In the case of fifth order methods, the eight-stage method has a larger effective SSP coefficient than the 12-stage method, so the eight stage method seems best in terms of efficiency. However the eight-stage method requires more storage registers (6) than the 12-stage method (5). So while the eight-stage method might be preferred for efficiency, the 12-stage method is preferred for low storage considerations.

Table 9.2 Effective SSP coefficients $\mathcal{C}_{\text{eff}}$ of optimal explicit two-step Runge–Kutta methods of order five to eight.

s \ p	5	6	7	8
4	0.214			
5	0.324			
6	0.385	0.099		
7	0.418	0.182		
8	0.447	0.242	0.071	
9	0.438	0.287	0.124	
10	0.425	0.320	0.179	
11	0.431	0.338	0.218	0.031
12	0.439	0.365	0.231	0.078

9.4 Coefficients of optimal methods

9.4.1 *Fifth order SSP TSRK methods*

9.4.1.1 *Coefficients of the optimal explicit Type II 5-stage 5th order SSP TSRK method*

$\tilde{\theta} = 0.02023608$

$\tilde{d}_2 = 0.111755628994$

$\tilde{d}_3 = 0.063774568581$

$\tilde{d}_5 = 0.061688513154$

$q_{2,0} = 0.165966644335$

$q_{2,1} = 0.722277726671$

$q_{3,0} = 0.130982771813$

$q_{3,2} = 0.805242659606$

$q_{4,1} = 0.233632131835$

$q_{4,3} = 0.476948026747$

$q_{5,1} = 0.533135921858$

$q_{5,4} = 0.375919004282$

$\eta_0 = 0.031579186622$

$\eta_1 = 0.237251868533$

$\eta_5 = 0.638227991821$

9.4.1.2 *Coefficients of the optimal explicit Type II 8-stage 5th order SSP TSRK method*

$\tilde{\theta} = 0.00000000$

$\tilde{d}_7 = 0.003674184820$

$q_{2,0} = 0.085330772948$

$q_{2,1} = 0.914669227052$

$q_{3,0} = 0.058121281984$

$q_{3,2} = 0.941878718016$

$q_{4,1} = 0.036365639243$

$q_{4,3} = 0.802870131353$

$q_{5,1} = 0.491214340661$

$q_{5,4} = 0.508785659339$

$q_{6,1} = 0.566135231631$

$q_{6,5} = 0.433864768369$

$q_{7,0} = 0.020705281787$

$q_{7,1} = 0.091646079652$

$q_{7,6} = 0.883974453742$

$q_{8,0} = 0.008506650139$

$q_{8,1} = 0.110261531523$

$q_{8,2} = 0.030113037742$

$q_{8,7} = 0.851118780596$

$\eta_2 = 0.179502832155$

$\eta_3 = 0.073789956885$

$\eta_6 = 0.017607159013$

$\eta_8 = 0.729100051947$

9.4.1.3 Coefficients of the optimal explicit Type II 12-stage 5th order SSP TSRK method

$\tilde{\theta} = 0.00000000$

$q_{2,0} = 0.037442206073$

$q_{2,1} = 0.962557793927$

$q_{3,0} = 0.004990369160$

$q_{3,2} = 0.750941165462$

$q_{4,3} = 0.816192058726$

$q_{5,4} = 0.881400968167$

$q_{6,1} = 0.041456384663$

$q_{6,5} = 0.897622496600$

$q_{7,1} = 0.893102584263$

$q_{7,6} = 0.106897415737$

$q_{8,6} = 0.197331844351$

$q_{8,7} = 0.748110262498$

$q_{9,1} = 0.103110842229$

$q_{9,8} = 0.864072067201$

$q_{10,1} = 0.109219062396$

$q_{10,9} = 0.890780937604$

$q_{11,1} = 0.069771767767$

$q_{11,10} = 0.928630488245$

$q_{12,1} = 0.050213434904$

$q_{12,11} = 0.949786565096$

$\eta_1 = 0.010869478270$

$\eta_6 = 0.252584630618$

$\eta_{10} = 0.328029300817$

$\eta_{12} = 0.408516590295$

9.4.2 Sixth order SSP TSRK methods

9.4.2.1 Coefficients of the optimal explicit Type II 8-stage 6th order SSP TSRK method

$\tilde{\theta} = 0.00106116$

$\tilde{d}_2 = 0.069865165650$

$\tilde{d}_5 = 0.031664390591$

$\tilde{d}_7 = 0.005723003763$

$q_{2,0} = 0.155163881254$

$q_{2,1} = 0.774970953096$

$q_{3,0} = 0.011527680249$

$q_{3,2} = 0.565240838867$

$q_{4,1} = 0.183055691416$

$q_{4,3} = 0.482238589858$

$q_{5,1} = 0.527099910423$

$q_{5,4} = 0.202048160624$

$q_{6,1} = 0.132151075381$

$q_{6,5} = 0.569109812305$

$q_{7,1} = 0.196585762013$

$q_{7,6} = 0.621546092130$

$q_{8,1} = 0.163546177998$

$q_{8,7} = 0.713116793634$

$\eta_1 = 0.082551065859$

$\eta_3 = 0.035081156057$

$\eta_4 = 0.148953483941$

$\eta_6 = 0.294405673499$

$\eta_7 = 0.137493811494$

$\eta_8 = 0.162571162883$

9.4.2.2 Coefficients of the optimal explicit Type II 10-stage 6th order SSP TSRK method

$\tilde{\theta} = 0.00036653$

$\tilde{d}_6 = 0.009912533094$

$\tilde{d}_7 = 0.002334846069$

$\tilde{d}_9 = 0.000530458859$

$q_{2,0} = 0.094376547634$

$q_{2,1} = 0.809292150916$

$q_{3,2} = 0.610221440928$

$q_{4,1} = 0.007667153484$

$q_{4,3} = 0.736481210931$

$q_{5,1} = 0.424854291345$

$q_{5,4} = 0.419583041927$

$q_{6,1} = 0.532557860710$

$q_{6,5} = 0.358181059904$

$q_{7,1} = 0.138195575844$

$q_{7,6} = 0.715575495505$

$q_{8,1} = 0.042985200527$

$q_{8,3} = 0.053034932664$

$q_{8,4} = 0.042145808612$

$q_{8,7} = 0.748638417763$

$q_{9,1} = 0.133834554661$

$q_{9,8} = 0.832560135198$

$q_{10,1} = 0.107245009824$

$q_{10,4} = 0.344370117388$

$q_{10,9} = 0.548384872788$

$\eta_1 = 0.037846106485$

$\eta_5 = 0.182774592259$

$\eta_7 = 0.001910649755$

$\eta_8 = 0.444471969041$

$\eta_{10} = 0.293438249715$

9.4.2.3 Coefficients of the optimal explicit Type II 12-stage 6th order SSP TSRK method

$\tilde{\theta} = 0.00024559$

$\tilde{d}_{10} = 0.000534877910$

$q_{2,0} = 0.030262100443$

$q_{2,1} = 0.664746114331$

$q_{3,2} = 0.590319496201$

$q_{4,3} = 0.729376762034$

$q_{5,4} = 0.826687833242$

$q_{6,1} = 0.656374628866$

$q_{6,5} = 0.267480130554$

$q_{7,1} = 0.210836921275$

$q_{7,6} = 0.650991182223$

$q_{8,7} = 0.873267220579$

$q_{9,1} = 0.066235890301$

$q_{9,8} = 0.877348047199$

$q_{10,1} = 0.076611491217$

$q_{10,4} = 0.091956261008$

$q_{10,9} = 0.822483564558$

$q_{11,4} = 0.135742974049$

$q_{11,5} = 0.269086406274$

$q_{11,10} = 0.587217894187$

$q_{12,1} = 0.016496364995$

$q_{12,5} = 0.344231433411$

$q_{12,6} = 0.017516154376$

$q_{12,11} = 0.621756047217$

$\eta_1 = 0.012523410806$

$\eta_6 = 0.094203091821$

$\eta_9 = 0.318700620500$

$\eta_{10} = 0.107955864652$

$\eta_{12} = 0.456039783327$

9.4.3 Seventh order SSP TSRK methods

9.4.3.1 Coefficients of the optimal explicit Type II 8-stage 7th order SSP TSRK method

$\tilde{\theta} = 0.00332617$

$\tilde{d}_2 = 0.126653296726$

$\tilde{d}_4 = 0.009857905982$

$\tilde{d}_6 = 0.010592856459$

$\tilde{d}_7 = 0.048409508995$

$\tilde{d}_8 = 0.002280862470$

$q_{2,0} = 0.031799261055$

$q_{2,1} = 0.179746357536$

$q_{3,1} = 0.039797621783$

$q_{3,2} = 0.124239161110$

$q_{4,1} = 0.096889072022$

$q_{4,3} = 0.078627093576$

$q_{5,0} = 0.000051639600$

$q_{5,1} = 0.055681474734$

$q_{5,4} = 0.180135981178$

$q_{6,1} = 0.102147320995$

$q_{6,5} = 0.210214808939$

$q_{7,0} = 0.008592669726$

$q_{7,1} = 0.141040850109$

$q_{7,2} = 0.010491388879$

$q_{7,6} = 0.261055198396$

$q_{8,1} = 0.062789906516$

$q_{8,2} = 0.008671822429$

$q_{8,5} = 0.241535951878$

$q_{8,7} = 0.080913601229$

$\eta_0 = 0.000438501336$

$\eta_1 = 0.052241502367$

$\eta_3 = 0.028268547821$

$\eta_4 = 0.035169559605$

$\eta_5 = 0.160342270301$

$\eta_8 = 0.158598041362$

9.4.3.2 Coefficients of the optimal explicit Type II 10-stage 7th order SSP TSRK method

$\tilde{\theta} = 0.00133831$

$\tilde{d}_2 = 0.060393526751$

$\tilde{d}_4 = 0.008021351370$

$\tilde{d}_5 = 0.000110261588$

$\tilde{d}_7 = 0.020315411877$

$\tilde{d}_{10} = 0.000798257806$

$q_{2,0} = 0.104220534835$

$q_{2,1} = 0.580488095534$

$q_{3,1} = 0.096961276680$

$q_{3,2} = 0.328772295105$

$q_{4,1} = 0.270102282192$

$q_{4,3} = 0.127716555038$

$q_{5,1} = 0.108198887289$

$q_{5,4} = 0.363940880582$

$q_{6,1} = 0.126367681916$

$q_{6,5} = 0.459633702775$

$q_{7,1} = 0.348220109451$

$q_{7,6} = 0.359028856085$

$q_{8,1} = 0.115894937032$

$q_{8,2} = 0.072893942403$

$q_{8,7} = 0.537696336776$

$q_{9,1} = 0.094490229781$

$q_{9,3} = 0.116506580533$

$q_{9,5} = 0.010233375426$

$q_{9,6} = 0.095165076447$

$q_{9,8} = 0.476926849074$

$q_{10,1} = 0.073865655852$

$q_{10,6} = 0.503612893798$

$q_{10,9} = 0.218719464641$

$\eta_0 = 0.000893197498$

$\eta_1 = 0.065663220513$

$\eta_2 = 0.000395829880$

$\eta_5 = 0.083001075910$

$\eta_6 = 0.272056402441$

$\eta_{10} = 0.417047864831$

9.4.3.3 Coefficients of the optimal explicit Type II 12-stage 7th order SSP TSRK method

$\tilde{\theta} = 0.00010402$

$\tilde{d}_2 = 0.003229110379$

$\tilde{d}_4 = 0.006337974350$

$\tilde{d}_5 = 0.002497954202$

$\tilde{d}_8 = 0.017328228771$

$\tilde{d}_{12} = 0.000520256251$

$q_{2,0} = 0.147321824258$

$q_{2,1} = 0.849449065363$

$q_{3,1} = 0.120943274105$

$q_{3,2} = 0.433019948758$

$q_{4,1} = 0.368587879162$

$q_{4,3} = 0.166320497215$

$q_{5,1} = 0.222052624372$

$q_{5,4} = 0.343703780759$

$q_{6,1} = 0.137403913799$

$q_{6,5} = 0.519758489994$

$q_{7,1} = 0.146278214691$

$q_{7,2} = 0.014863996842$

$q_{7,6} = 0.598177722196$

$q_{8,1} = 0.444640119039$

$q_{8,7} = 0.488244475585$

$q_{9,1} = 0.143808624107$

$q_{9,2} = 0.026942009774$

$q_{9,8} = 0.704865150213$

$q_{10,1} = 0.102844296820$

$q_{10,3} = 0.032851385162$

$q_{10,7} = 0.356898323452$

$q_{10,9} = 0.409241038172$

$q_{11,1} = 0.071911085489$

$q_{11,7} = 0.508453150788$

$q_{11,10} = 0.327005955933$

$q_{12,1} = 0.057306282669$

$q_{12,7} = 0.496859299070$

$q_{12,11} = 0.364647377607$

$\eta_0 = 0.000515717568$

$\eta_1 = 0.040472655980$

$\eta_6 = 0.081167924336$

$\eta_7 = 0.238308176460$

$\eta_8 = 0.032690786324$

$\eta_{12} = 0.547467490509$

9.4.4 Eighth order SSP TSRK methods

9.4.4.1 Coefficients of the optimal explicit Type II 12-stage 8th order SSP TSRK method

$\tilde{\theta} = 0.00004796$

$\tilde{d}_2 = 0.036513886686$

$\tilde{d}_4 = 0.004205435886$

$\tilde{d}_5 = 0.000457751617$

$\tilde{d}_7 = 0.007407526544$

$\tilde{d}_8 = 0.000486094554$

$q_{2,0} = 0.017683145597$

$q_{2,1} = 0.154785324943$

$q_{3,0} = 0.001154189099$

$q_{3,2} = 0.200161251442$

$q_{4,1} = 0.113729301017$

$q_{4,3} = 0.057780552515$

$q_{5,1} = 0.061188134341$

$q_{5,4} = 0.165254103192$

$q_{6,0} = 0.000065395820$

$q_{6,1} = 0.068824803789$

$q_{6,2} = 0.008642531617$

$q_{6,5} = 0.229847794525$

$q_{7,1} = 0.133098034326$

$q_{7,4} = 0.005039627904$

$q_{7,6} = 0.252990567223$

$q_{8,1} = 0.080582670157$

$q_{8,4} = 0.069726774932$

$q_{8,7} = 0.324486261337$

$q_{9,0} = 0.000042696256$

$q_{9,1} = 0.038242841052$

$q_{9,3} = 0.029907847390$

$q_{9,4} = 0.022904196668$

$q_{9,5} = 0.095367316002$

$q_{9,6} = 0.176462398918$

$q_{9,8} = 0.120659479468$

$q_{10,1} = 0.071728403471$

$q_{10,6} = 0.281349762795$

$q_{10,9} = 0.166819833905$

$q_{11,0} = 0.000116117870$

$q_{11,1} = 0.053869626312$

$q_{11,6} = 0.327578464732$

$q_{11,10} = 0.157699899496$

$q_{12,0} = 0.000019430721$

$q_{12,1} = 0.009079504343$

$q_{12,4} = 0.130730221737$

$q_{12,6} = 0.149446805276$

$q_{12,11} = 0.314802533082$

$\eta_1 = 0.033190060418$

$\eta_2 = 0.001567085178$

$\eta_3 = 0.014033053075$

$\eta_4 = 0.017979737867$

$\eta_5 = 0.094582502433$

$\eta_6 = 0.082918042281$

$\eta_7 = 0.020622633348$

$\eta_8 = 0.033521998905$

$\eta_9 = 0.092066893963$

$\eta_{10} = 0.076089630105$

$\eta_{11} = 0.070505470986$

$\eta_{12} = 0.072975312278$

Chapter 10

Downwinding

10.1 SSP methods with negative β_{ij}'s

As we have seen in the previous chapters, SSP Runge–Kutta methods suffer from order barriers and tight bounds on the SSP coefficient. For many popular Runge–Kutta methods, including the classical fourth order method, it turns out that the SSP coefficient $\mathcal{C} = 0$, so that the method is not SSP under any positive time step.

We have seen that Runge–Kutta methods with large SSP coefficients can be obtained by using extra stages (Chapter 6) and methods with higher order accuracy can be obtained by using additional steps (Chapters 8 and 9). In this chapter we describe another approach for obtaining methods with higher order and larger SSP coefficient.

From the very first design of SSP methods in [91, 92], downwinding has been suggested to expand the scope of applicability of SSP methods (see Remark 2.1). It is much easier if we insist only on the positivity of α_{ij}'s in the Shu-Osher form (2.10) in Chapter 2, while allowing β_{ij}'s in (2.10) to become negative. Of course, for such time discretizations, the assumption of strong stability for the forward Euler step (2.6) alone is not enough to guarantee SSP for the high order time discretization (2.10).

However, in the solution of hyperbolic conservation laws, the SSP property can also be guaranteed for such methods, provided that we use a modified spatial discretization for these instances. The semi-discretization F of a hyperbolic system typically involves some form of upwinding, representing numerical viscosity. Therefore, if F is a discretization to $-f(u)_x$ for the PDE

$$u_t + f(u)_x = 0 \tag{10.1}$$

satisfying the strong stability property

$$\|u^n + \Delta t F(u^n)\| \le \|u^n\| \tag{10.2}$$

under the time step restriction

$$\Delta t \le \Delta t_{\text{FE}}, \tag{10.3}$$

then typically the same F used to approximate $-f(u)_x$ in the PDE

$$u_t - f(u)_x = 0 \tag{10.4}$$

would generate an unstable scheme

$$u^{n+1} = u^n - \Delta t F(u^n).$$

However, we note that (10.4) is also a hyperbolic conservation law, the only difference with (10.1) being that the flux $f(u)$ is replaced by $-f(u)$. Since we can build a stable spatial discretization F for the PDE (10.1) to satisfy the strong stability condition (10.2) under the time step restriction (10.3), the same procedure should allow us to build another stable spatial discretization, denoted by $-\tilde{F}$, which approximates $f(u)_x$ in the PDE (10.4) to generate a scheme

$$u^{n+1} = u^n - \Delta t \tilde{F}(u^n) \tag{10.5}$$

which is strongly stable

$$\|u^n - \Delta t \tilde{F}(u^n)\| \le \|u^n\| \tag{10.6}$$

under the same time step restriction (10.3). Typically, this only involves a different direction for upwinding. That is, if the wind direction for (10.1) is from left to right, hence the stable approximation F in (10.2) has a left-biased stencil, then the wind direction for (10.4) is from right to left, hence the stable approximation $-\tilde{F}$ in (10.6) would have a right-biased stencil. Notice that both F and $\tilde{F}$ are approximations to the same spatial operator $f(u)_x$ with the same order of accuracy, their only difference is in the choice of upwinding direction. $\tilde{F}$ is often referred to as the downwinding approximation to $-f(u)_x$. For instance, if in the PDE (10.1) we have $f'(u) \ge 0$, hence the wind direction is from left to right, then the first order upwind scheme

$$u^{n+1} = u^n + \Delta t F(u^n)$$

with

$$F(u)_j = -\frac{1}{\Delta x}\left(f(u_j) - f(u_{j-1})\right)$$

is strongly stable under the time step restriction $\Delta t \leq \Delta t_{\text{FE}}$ with $\Delta t_{\text{FE}} = \frac{\Delta x}{\max_u |f'(u)|}$. Then, the first order downwind scheme

$$u^{n+1} = u^n - \Delta t \tilde{F}(u^n)$$

with

$$\tilde{F}(u)_j = -\frac{1}{\Delta x} \left(f(u_{j+1}) - f(u_j) \right)$$

is actually a stable upwind scheme for the conservation law (10.4) under the same time step restriction. Similar construction can be generalized to higher order linear finite difference discretizations, the matrix representing $\tilde{F}$ being just the transpose of that representing F. For all other familiar high resolution schemes, such as TVD, ENO and WENO finite difference or finite volume schemes and discontinuous Galerkin finite element schemes, the discretization $\tilde{F}$ can be constructed in a similarly straightforward way.

A general m-stage, explicit Runge–Kutta method is written in the Shu-Osher form [92]:

$$u^{(0)} = u^n$$

$$u^{(i)} = \sum_{j=0}^{i-1} \alpha_{ij} u^{(k)} + \Delta t \beta_{ij} \begin{cases} F(u^{(j)}) & \text{if } \beta_{ij} \geq 0 \\ \tilde{F}(u^{(j)}) & \text{otherwise} \end{cases}, \qquad i = 1, 2, \ldots, m,$$

$$u^{n+1} = u^{(m)}.$$

That is, $F(u^{(j)})$ is replaced by $\tilde{F}(u^{(j)})$ whenever the corresponding coefficient β_{ij} is negative.

If all the $\alpha_{ij} \geq 0$, then this form is a convex combination of explicit Euler steps, which allows us to prove that the solution obtained by this method satisfies the strong stability bound

$$\|u^{n+1}\| \leq \|u^n\|$$

under the time step restriction

$$\Delta t \leq \tilde{C} \Delta t_{FE}, \tag{10.7}$$

where $\tilde{C} = \min_{i,j} \frac{\alpha_{ij}}{|\beta_{ij}|}$, and the ratio is understood as infinite if $\beta_{ij} = 0$.

Similarly, an explicit linear multistep method is written as a convex combination of explicit Euler steps:

$$u^{n+1} = \sum_{i=1}^{s} \left(\alpha_i u^{n+1-i} + \Delta t \beta_i \begin{cases} F(u^{n+1-i}) & \text{if } \beta_i \geq 0 \\ \tilde{F}(u^{n+1-i}) & \text{otherwise} \end{cases} \right)$$

and can be shown, therefore, to be SSP under the step size restriction $\Delta t \leq \tilde{C} \Delta t_{FE}$, with $\tilde{C} = \min_i \frac{\alpha_i}{|\beta_i|}$. As before, we refer to $\tilde{C}$ as the *SSP coefficient*, and define the *effective SSP coefficient* $\tilde{C}_{\text{eff}}$ to be the SSP coefficient divided by the number of function evaluations.

It would seem that if both F and $\tilde{F}$ must be computed for the same stage (or step), the computational cost as well as the storage requirement for this stage is doubled. For this reason, negative coefficients were avoided whenever possible in [26–28, 86, 98]. However, using downwinding we can obtain SSP methods with positive $\tilde{C}$ even in classes where all methods have $C = 0$ (see, e.g. Proposition 3.3 of [26] and Theorem 4.1 in [86]). For example, a four-stage fourth order method with downwinding can have a positive SSP coefficient $\tilde{C}$ even though any four-stage fourth order method has $C = 0$. For this reason, recent studies (e.g. [87, 84, 30]) have considered efficient ways of implementing downwind discretizations. Inclusion of negative coefficients, may raise the SSP coefficient enough to compensate for the additional computational cost incurred by $\tilde{F}$. Since $\tilde{F}$ is, numerically, the downwind version of F, it is sometimes possible to compute both F and $\tilde{F}$ without doubling the computational cost [30]. If F and $\tilde{F}$ do not appear for the same stage, then neither the computational cost nor the storage requirement is increased.

10.2 Explicit SSP Runge–Kutta methods with downwinding

In this section we give examples of explicit SSP Runge–Kutta methods with downwinding. In this section assume that the cost of evaluating both F and $\tilde{F}$ is double that of computing F alone. However, if we assume that the cost of evaluating both operators is $1 + \delta$ general function evaluations where $0 \leq \delta \leq 1$, each step of the algorithm will only have a cost of $1 + \delta$ general function evaluations if any downwind operators are present. For more details on SSP methods with downwinding, see [20, 27, 29, 30, 87, 84, 92, 91, 96, 71]. For an alternative perspective on downwinded methods as additive discretizations, see [37–39]. For some upper bounds on the SSP coefficient of explicit general linear methods with downwinding, see [57].

10.2.1 *Second and third order methods*

As we saw in Chapter 2, excellent SSP methods exist in the class of explicit second and third order methods without downwinding. However, downwinding allows a higher SSP coefficient, even if we assume that the cost to compute both F and $\tilde{F}$ for a given stage is double that of computing F alone.

Explicit SSP Runge–Kutta with downwinding SSPRK$_*$(2,2): The two-stage second order method has SSP coefficient $\tilde{C} = 1.2152504$. It requires three function evaluations per step, for an effective SSP coefficient $\tilde{C}_{\text{eff}} = 0.405083467$. This method is given by coefficients:

$$\alpha_{10} = 1.000000000000000 \quad \beta_{10} = 0.822875655532364$$
$$\alpha_{20} = 0.261583187659478 \quad \beta_{20} = -0.215250437021539$$
$$\alpha_{21} = 0.738416812340522 \quad \beta_{21} = 0.607625218510713$$

Explicit SSP Runge–Kutta with downwinding SSPRK$_*$(3,2): The three-stage second order method has SSP coefficient $\tilde{C} = 2.1861407$. It requires four function evaluations per step, for an effective SSP coefficient $\tilde{C}_{\text{eff}} = 0.546535175$. This method is given by coefficients:

$$\alpha_{10} = 1.000000000000000 \quad \beta_{10} = 0.457427107756303$$
$$\alpha_{20} = 0.000000000000000 \quad \beta_{20} = 0.000000000000000$$
$$\alpha_{21} = 1.000000000000000 \quad \beta_{21} = 0.457427107756303$$
$$\alpha_{30} = 0.203464834591289 \quad \beta_{30} = -0.093070330817223$$
$$\alpha_{31} = 0.000000000000000 \quad \beta_{31} = 0.000000000000000$$
$$\alpha_{32} = 0.796535165408711 \quad \beta_{32} = 0.364356776939073$$

Explicit SSP Runge–Kutta with downwinding SSPRK$_*$(3,3): The three-stage third order method has SSP coefficient $\tilde{C} = 1.3027756$. However, it requires four function evaluations per step, for an effective SSP coefficient $\tilde{C}_{\text{eff}} = 0.3256939$ if $\delta = 1$, which is less than that of the SSPRK(3,3) method without downwinding. If the cost of extra evaluations is small, this method may be more efficient. This method is given by coefficients:

$$\alpha_{10} = 1.000000000000000 \quad \beta_{10} = 0.767591879243998$$
$$\alpha_{20} = 0.410802706918667 \quad \beta_{20} = -0.315328821802221$$
$$\alpha_{21} = 0.589197293081333 \quad \beta_{21} = 0.452263057441777$$
$$\alpha_{30} = 0.123062611901395 \quad \beta_{30} = -0.041647109531262$$
$$\alpha_{31} = 0.251481201947289 \quad \beta_{31} = 0.000000000000000$$
$$\alpha_{32} = 0.625456186151316 \quad \beta_{32} = 0.480095089312672$$

10.2.2 Fourth order methods

We mentioned before that all four-stage, fourth order Runge–Kutta methods with positive SSP coefficient $\tilde{C}$ must have at least one negative β_{ij}; see [62, 26]. Therefore, we can add another stage to look for five-stage, fourth order explicit SSP Runge–Kutta methods with all positive coefficients, with one such example SSPRK(5,4) described in Chapter 2 having an effective SSP coefficient $\tilde{C}_{\text{eff}} = 0.302$. Alternatively, we can find a four-stage, fourth order Runge–Kutta method with some negative β_{ij}. One such example is given in [26]:

$$u^{(1)} = u^n + \frac{1}{2}\Delta t F(u^n)$$

$$u^{(2)} = \frac{649}{1600}u^{(0)} - \frac{10890423}{25193600}\Delta t \tilde{F}(u^n) + \frac{951}{1600}u^{(1)} + \frac{5000}{7873}\Delta t F(u^{(1)})$$

$$u^{(3)} = \frac{53989}{2500000}u^n - \frac{102261}{5000000}\Delta t \tilde{F}(u^n) + \frac{4806213}{20000000}u^{(1)} \qquad (10.8)$$
$$\quad - \frac{5121}{20000}\Delta t \tilde{F}(u^{(1)}) + \frac{23619}{32000}u^{(2)} + \frac{7873}{10000}\Delta t F(u^{(2)})$$

$$u^{n+1} = \frac{1}{5}u^n + \frac{1}{10}\Delta t F(u^n) + \frac{6127}{30000}u^{(1)} + \frac{1}{6}\Delta t F(u^{(1)}) + \frac{7873}{30000}u^{(2)}$$
$$\quad + \frac{1}{3}u^{(3)} + \frac{1}{6}\Delta t F(u^{(3)}).$$

This method is fourth order accurate with a SSP coefficient $\tilde{C} = 0.936$. Notice that two $\tilde{F}$'s must be computed, hence the effective SSP coefficient under the assumption $\delta = 1$ is $\tilde{C}_{\text{eff}} = 0.936/6 = 0.156$. This is much less efficient than the SSPRK(5,4) method described in Chapter 2 which has an effective SSP coefficient $\mathcal{C}_{\text{eff}} = 0.302$.

A slightly better method is given by coefficients

$$\alpha_{10} = 1.000000000000000 \qquad \beta_{10} = 0.545797148202810$$

$$\alpha_{20} = 0.447703597093315 \qquad \beta_{20} = -0.455917323951788$$

$$\alpha_{21} = 0.552296402906685 \qquad \beta_{21} = 0.562429025981069$$

$$\alpha_{30} = 0.174381001639320 \qquad \beta_{30} = -0.177580256517037$$

$$\alpha_{31} = 0.000000000000000 \qquad \beta_{31} = 0.000000000000000$$

$$\alpha_{32} = 0.825618998360680 \qquad \beta_{32} = 0.840766093415820$$

$$\alpha_{40} = 0.374455263824577 \qquad \beta_{40} = 0.107821590754283$$

$$\alpha_{41} = 0.271670479800689 \qquad \beta_{41} = 0.276654641489540$$

$$\alpha_{42} = 0.081190815217391 \qquad \beta_{42} = 0.000000000000000$$

$$\alpha_{43} = 0.272683441157343 \qquad \beta_{43} = 0.161441275936663.$$

This method has $\tilde{C} = 0.9819842$ and requires five function evaluations, for an effective SSP coefficient $\tilde{C}_{\text{eff}} = 0.9819842/(4+\delta) = 0.19639684$ for $\delta = 1$.

Adding another stage as well as allowing downwinding allows an even larger SSP coefficient of $\tilde{C} = 2.0312031$

$$\alpha_{10} = 1.000000000000000 \quad \beta_{10} = 0.416596471458169$$
$$\alpha_{20} = 0.210186660827794 \quad \beta_{20} = -0.103478898431154$$
$$\alpha_{21} = 0.789813339172206 \quad \beta_{21} = 0.388840157514713$$
$$\alpha_{30} = 0.331062996240662 \quad \beta_{30} = -0.162988621767813$$
$$\alpha_{31} = 0.202036516631465 \quad \beta_{31} = 0.000000000000000$$
$$\alpha_{32} = 0.466900487127873 \quad \beta_{32} = 0.229864007043460$$
$$\alpha_{40} = 0.000000000000000 \quad \beta_{40} = 0.000000000000000$$
$$\alpha_{41} = 0.000000000000000 \quad \beta_{41} = 0.000000000000000$$
$$\alpha_{42} = 0.000000000000000 \quad \beta_{42} = 0.000000000000000$$
$$\alpha_{43} = 1.000000000000000 \quad \beta_{43} = 0.492319055945867$$
$$\alpha_{50} = 0.097315407775058 \quad \beta_{50} = -0.047910229684804$$
$$\alpha_{51} = 0.435703937692290 \quad \beta_{51} = 0.202097732052527$$
$$\alpha_{52} = 0.000000000000000 \quad \beta_{52} = 0.000000000000000$$
$$\alpha_{53} = 0.000000000000000 \quad \beta_{53} = 0.000000000000000$$
$$\alpha_{54} = 0.466980654532652 \quad \beta_{54} = 0.229903474984498$$

This method requires six function evaluations, for an effective SSP coefficient of $\tilde{C}_{\text{eff}} = 2.0312031/(5 + \delta) = 0.33853385$ if $\delta = 1$.

10.2.3 A fifth order method

We saw in Chapter 5 that there is no SSP fifth order Runge–Kutta method with non-negative coefficients, regardless of how many stages one would like to use. Therefore, we must consider SSP Runge–Kutta methods with some negative β_{ij}'s for fifth order methods. In [87], Ruuth and Spiteri constructed explicit fifth order SSP Runge–Kutta methods with some negative β_{ij}'s for fifth order and seven, eight or nine stages. These schemes were found by a numerical search to optimize the effective SSP coefficient. Their seven-stage, fifth order scheme is given by

$$K_i = \begin{cases} F\left(u^n + \Delta t \sum_{j=1}^{i-1} a_{ij} K_j\right) & \text{if } b_i \geq 0, \\ \tilde{F}\left(u^n + \Delta t \sum_{j=1}^{i-1} a_{ij} K_j\right) & \text{otherwise,} \end{cases} \qquad i = 1, 2, \cdots, 7$$

$$u^{n+1} = u^n + \Delta t \sum_{i=1}^{7} b_i K_i$$

with the following coefficients

$$
\begin{aligned}
a_{2,1} &= 0.392382208054010, & a_{3,1} &= 0.310348765296963, \\
a_{3,2} &= 0.523846724909595, & a_{4,1} &= 0.114817342432177, \\
a_{4,2} &= 0.248293597111781, & a_{4,3} &= 0, \\
a_{5,1} &= 0.136041285050893, & a_{5,2} &= 0.163250087363657, \\
a_{5,3} &= 0, & a_{5,4} &= 0.557898557725281, \\
a_{6,1} &= 0.135252145083336, & a_{6,2} &= 0.207274083097540, \\
a_{6,3} &= -0.180995372278096, & a_{6,4} &= 0.326486467604174, \\
a_{6,5} &= 0.348595427190109, & a_{7,1} &= 0.082675687408986, \\
a_{7,2} &= 0.146472328858960, & a_{7,3} &= -0.160507707995237, \\
a_{7,4} &= 0.161924299217425, & a_{7,5} &= 0.028864227879979, \\
a_{7,6} &= 0.070259587451358, & b_1 &= 0.110184169931401, \\
b_2 &= 0.122082833871843, & b_3 &= -0.117309105328437, \\
b_4 &= 0.169714358772186, & b_5 &= 0.143346980044187, \\
b_6 &= 0.348926696469455, & b_7 &= 0.223054066239366.
\end{aligned}
$$

This method is fifth order accurate with a SSP coefficient $\tilde{C} = 1.1785$. Since there are seven evaluations of F or $\tilde{F}$, the effective SSP coefficient is $\tilde{C}_{\text{eff}} = 1.1785/7 = 0.168$. For the coefficients of eight- and nine-stage fifth order SSP Runge–Kutta methods, we refer the reader to [86].

10.3 Optimal explicit multistep schemes with downwinding

It can be shown that the overall upper bounds on the threshold factor for multistep methods with downwinding are the same as for methods without downwinding [57]. So for explicit methods, the SSP coefficient is bounded by $\tilde{C} \leq 1$ and for implicit methods $\tilde{C} \leq 2$. Fortunately, the bound (8.9) [67] does *not* hold for methods with downwinding.

Optimal s-step, order p multistep methods with downwinding were found in [91, 27, 85] for $s \leq 6, 2 \leq p \leq 6$ and $s \geq 2, p = 2$. These methods were found again in [30] and guaranteed optimal using the deterministic global branch-and-bound software, BARON [88]. Using a variation of the optimization technique described Chapter 4, explicit and implicit methods of up to $p = 15$ and $p = 50$ were found in [56, 57]. We present their SSP coefficients in Tables 10.1 and 10.2. We see from these tables that downwinding often leads to improved effective SSP coefficients. This is especially true for methods of high order that do not use too many steps. For instance, constructing a tenth order explicit SSP linear multistep method with downwinding requires only ten stages, whereas without downwinding such methods must have at least 22 stages.

One might assume that the cost of evaluating both F and $\tilde{F}$ is double that of computing F alone. However, if we assume that the cost of evaluating both operators is $1 + \delta$ general function evaluations where $0 \leq \delta \leq 1$, each step of the algorithm will only have a cost of $1 + \delta$ general function evaluations if any downwind operators are present. Optimal methods for this situation were designed in [30], where the size of the increment, δ, determines which scheme is more efficient.

Explicit SSPMS$_*$(7,5) The seven-step fifth order multistep method with downwinding has $\tilde{C} = 0.1868460$, and effective SSP coefficient $\tilde{C}_{\text{eff}} = \frac{0.1868460}{1+\delta}$, and is defined by the coefficients:

$$\alpha_1 = 0.437478073273716 \qquad \beta_1 = 2.341383323503706$$
$$\alpha_2 = 0.177079742280077 \qquad \beta_2 = -0.947731054044159$$
$$\alpha_4 = 0.266879475710902 \qquad \beta_4 = 1.428339365991395$$
$$\alpha_6 = 0.079085404949912 \qquad \beta_6 = -0.423265209377492$$
$$\alpha_7 = 0.039477303785393 \qquad \beta_7 = 0.211282590801251$$

Explicit SSPMS$_*$(10,6) The ten-step sixth order multistep method with downwinding has $\tilde{C} = 0.1749490$, and effective SSP coefficient $\tilde{C}_{\text{eff}} = \frac{0.1749490}{1+\delta}$, and is defined by the coefficients:

$$\alpha_1 = 0.421496355190108 \qquad \beta_1 = 2.409253340733589$$
$$\alpha_2 = 0.184871618144855 \qquad \beta_2 = -1.056717473684455$$
$$\alpha_4 = 0.261496145095487 \qquad \beta_4 = 1.494699665620621$$
$$\alpha_7 = 0.030002986393737 \qquad \beta_7 = -0.171495658990894$$
$$\alpha_9 = 0.078557623043187 \qquad \beta_9 = 0.449031678275387$$
$$\alpha_{10} = 0.023575272132626 \qquad \beta_{10} = -0.134755146621380$$

Table 10.1 SSP coefficients for optimal s-step, pth-order linear multistep schemes with downwinding

	1	2	3	4	5	6	7	8	9	10	11	12	13	14	15
1	1.000														
2	1.000	0.500													
3	1.000	0.667	0.287												
4	1.000	0.750	0.415	0.159											
5	1.000	0.800	0.517	0.237	0.087										
6	1.000	0.833	0.583	0.283	0.131	0.046									
7	1.000	0.857	0.583	0.360	0.187	0.081	0.024								
8	1.000	0.875	0.583	0.394	0.223	0.107	0.044	0.013							
9	1.000	0.889	0.583	0.424	0.261	0.142	0.068	0.025	0.007						
10	1.000	0.900	0.583	0.447	0.299	0.175	0.086	0.037	0.013	0.003					
11	1.000	0.909	0.583	0.464	0.326	0.199	0.104	0.056	0.022	0.007	0.002				
12	1.000	0.917	0.583	0.477	0.351	0.215	0.131	0.069	0.030	0.012	0.004	0.000			
13	1.000	0.923	0.583	0.487	0.372	0.242	0.154	0.083	0.044	0.019	0.007	0.002	0.000		
14	1.000	0.929	0.583	0.496	0.391	0.266	0.171	0.096	0.055	0.024	0.012	0.004	0.001	0.000	
15	1.000	0.933	0.583	0.503	0.407	0.280	0.187	0.112	0.066	0.034	0.016	0.006	0.002	0.001	0.000
16	1.000	0.938	0.583	0.509	0.411	0.295	0.199	0.133	0.075	0.043	0.020	0.008	0.003	0.002	0.000
17	1.000	0.941	0.583	0.514	0.411	0.307	0.216	0.147	0.086	0.051	0.025	0.013	0.005	0.002	0.001
18	1.000	0.944	0.583	0.519	0.411	0.317	0.232	0.161	0.095	0.060	0.032	0.016	0.007	0.003	0.001
19	1.000	0.947	0.583	0.523	0.411	0.326	0.243	0.173	0.111	0.067	0.040	0.020	0.010	0.004	0.002
20	1.000	0.950	0.583	0.526	0.411	0.334	0.254	0.182	0.125	0.077	0.046	0.024	0.013	0.005	0.002
21	1.000	0.952	0.583	0.529	0.411	0.339	0.263	0.189	0.135	0.084	0.053	0.030	0.015	0.007	0.004
22	1.000	0.955	0.583	0.532	0.411	0.344	0.272	0.201	0.146	0.090	0.059	0.036	0.018	0.010	0.004
23	1.000	0.957	0.583	0.535	0.411	0.349	0.280	0.212	0.157	0.103	0.066	0.040	0.022	0.012	0.006
24	1.000	0.958	0.583	0.537	0.411	0.353	0.288	0.220	0.164	0.113	0.073	0.047	0.026	0.015	0.007
25	1.000	0.960	0.583	0.539	0.411	0.356	0.295	0.228	0.171	0.122	0.079	0.052	0.031	0.017	0.009
26	1.000	0.962	0.583	0.541	0.411	0.359	0.302	0.235	0.177	0.132	0.085	0.057	0.036	0.020	0.011
27	1.000	0.963	0.583	0.543	0.411	0.362	0.307	0.241	0.183	0.140	0.092	0.063	0.040	0.023	0.013
28	1.000	0.964	0.583	0.544	0.411	0.365	0.312	0.246	0.189	0.147	0.102	0.069	0.045	0.027	0.015
29	1.000	0.966	0.583	0.546	0.411	0.367	0.317	0.250	0.196	0.154	0.109	0.074	0.050	0.031	0.018
30	1.000	0.967	0.583	0.547	0.411	0.369	0.319	0.254	0.202	0.159	0.117	0.079	0.054	0.034	0.020
31	1.000	0.968	0.583	0.548	0.411	0.371	0.319	0.257	0.209	0.164	0.124	0.084	0.058	0.038	0.023
32	1.000	0.969	0.583	0.549	0.411	0.373	0.319	0.261	0.214	0.169	0.131	0.089	0.064	0.043	0.026
33	1.000	0.970	0.583	0.550	0.411	0.374	0.319	0.265	0.219	0.173	0.137	0.096	0.068	0.047	0.029
34	1.000	0.971	0.583	0.551	0.411	0.376	0.319	0.268	0.224	0.178	0.142	0.102	0.072	0.050	0.033
35	1.000	0.971	0.583	0.552	0.411	0.377	0.319	0.271	0.228	0.182	0.146	0.109	0.076	0.054	0.036
36	1.000	0.972	0.583	0.553	0.411	0.378	0.319	0.273	0.231	0.185	0.151	0.115	0.081	0.058	0.039
37	1.000	0.973	0.583	0.554	0.411	0.379	0.319	0.276	0.235	0.190	0.155	0.120	0.085	0.062	0.043
38	1.000	0.974	0.583	0.555	0.411	0.380	0.319	0.278	0.238	0.194	0.159	0.126	0.089	0.066	0.046
39	1.000	0.974	0.583	0.556	0.411	0.381	0.319	0.279	0.240	0.198	0.162	0.130	0.095	0.070	0.049
40	1.000	0.975	0.583	0.556	0.411	0.382	0.319	0.281	0.243	0.202	0.166	0.134	0.100	0.073	0.052

Table 10.2 Optimal SSP coefficients for implicit linear multistep methods with downwinding

	1	2	3	4	5	6	7	8	9	10	11	12	13	14	15
1	∞														
2	∞	2.000	1.303												
3	∞	2.000	1.591	1.000											
4	∞	2.000	1.710	1.243	0.744										
5	∞	2.000	1.776	1.243	0.871	0.544									
6	∞	2.000	1.817	1.243	0.976	0.672	0.384								
7	∞	2.000	1.845	1.243	1.035	0.788	0.535	0.261							
8	∞	2.000	1.866	1.243	1.073	0.870	0.619	0.410	0.171						
9	∞	2.000	1.882	1.243	1.093	0.905	0.668	0.474	0.280	0.107					
10	∞	2.000	1.894	1.243	1.117	0.905	0.718	0.542	0.369	0.191	0.065				
11	∞	2.000	1.905	1.243	1.132	0.905	0.755	0.585	0.433	0.280	0.119	0.038			
12	∞	2.000	1.913	1.243	1.143	0.905	0.776	0.629	0.485	0.304	0.195	0.078	0.021		
13	∞	2.000	1.920	1.243	1.153	0.905	0.791	0.664	0.512	0.376	0.234	0.121	0.047	0.013	
14	∞	2.000	1.926	1.243	1.160	0.905	0.803	0.693	0.540	0.424	0.296	0.175	0.077	0.028	0.007
15	∞	2.000	1.931	1.243	1.167	0.905	0.813	0.714	0.570	0.455	0.342	0.227	0.116	0.050	0.014
16	∞	2.000	1.935	1.243	1.172	0.905	0.821	0.719	0.588	0.479	0.376	0.240	0.157	0.074	0.029
17	∞	2.000	1.939	1.243	1.177	0.905	0.828	0.719	0.605	0.497	0.404	0.266	0.199	0.103	0.047
18	∞	2.000	1.943	1.243	1.181	0.905	0.834	0.719	0.620	0.520	0.430	0.330	0.230	0.140	0.067
19	∞	2.000	1.946	1.243	1.185	0.905	0.839	0.719	0.630	0.536	0.445	0.354	0.263	0.170	0.089
20	∞	2.000	1.949	1.243	1.188	0.905	0.843	0.719	0.638	0.551	0.457	0.379	0.292	0.201	0.124
21	∞	2.000	1.951	1.243	1.191	0.905	0.847	0.719	0.645	0.565	0.472	0.396	0.313	0.227	0.148
22	∞	2.000	1.953	1.243	1.194	0.905	0.850	0.719	0.651	0.577	0.484	0.413	0.335	0.256	0.172
23	∞	2.000	1.956	1.243	1.196	0.905	0.853	0.719	0.656	0.587	0.495	0.424	0.352	0.276	0.196
24	∞	2.000	1.957	1.243	1.198	0.905	0.856	0.719	0.660	0.595	0.504	0.435	0.369	0.293	0.220
25	∞	2.000	1.959	1.243	1.200	0.905	0.858	0.719	0.664	0.596	0.513	0.443	0.383	0.308	0.240
26	∞	2.000	1.961	1.243	1.202	0.905	0.860	0.719	0.668	0.596	0.520	0.453	0.392	0.325	0.258
27	∞	2.000	1.962	1.243	1.203	0.905	0.862	0.719	0.671	0.596	0.526	0.460	0.402	0.339	0.274
28	∞	2.000	1.964	1.243	1.205	0.905	0.864	0.719	0.674	0.596	0.532	0.469	0.410	0.352	0.289
29	∞	2.000	1.965	1.243	1.206	0.905	0.866	0.719	0.676	0.596	0.536	0.477	0.417	0.362	0.301
30	∞	2.000	1.966	1.243	1.208	0.905	0.867	0.719	0.678	0.596	0.540	0.483	0.423	0.370	0.314
31	∞	2.000	1.967	1.243	1.209	0.905	0.869	0.719	0.679	0.596	0.543	0.489	0.430	0.379	0.326
32	∞	2.000	1.968	1.243	1.210	0.905	0.870	0.719	0.681	0.596	0.547	0.495	0.435	0.386	0.336
33	∞	2.000	1.969	1.243	1.211	0.905	0.871	0.719	0.682	0.596	0.549	0.500	0.439	0.392	0.343
34	∞	2.000	1.970	1.243	1.212	0.905	0.872	0.719	0.684	0.596	0.552	0.505	0.444	0.398	0.350
35	∞	2.000	1.971	1.243	1.213	0.905	0.873	0.719	0.685	0.596	0.554	0.508	0.447	0.403	0.356
36	∞	2.000	1.972	1.243	1.214	0.905	0.874	0.719	0.686	0.596	0.556	0.508	0.451	0.408	0.363
37	∞	2.000	1.973	1.243	1.215	0.905	0.875	0.719	0.687	0.596	0.558	0.508	0.455	0.413	0.369
38	∞	2.000	1.973	1.243	1.216	0.905	0.876	0.719	0.688	0.596	0.559	0.508	0.458	0.417	0.374
39	∞	2.000	1.974	1.243	1.216	0.905	0.877	0.719	0.689	0.596	0.560	0.508	0.461	0.421	0.378
40	∞	2.000	1.975	1.243	1.217	0.905	0.878	0.719	0.690	0.596	0.562	0.508	0.463	0.425	0.382

10.4   Application: Deferred correction methods

In the previous sections we saw that using a downwind operator allows us to find SSP Runge–Kutta methods of higher order and larger SSP coefficients. A major area where downwinding can be useful is in the time stepping methods known as (spectral) deferred correction methods [18]. An advantage of the (spectral) deferred correction methods [18] is that they are one step methods, just like the general Runge–Kutta methods, but they can be constructed easily and systematically for any order of accuracy. This is in contrast to Runge–Kutta methods which are more difficult to construct for higher order of accuracy, and to multistep methods which need more storage space and are more difficult to restart with a different choice of the time step Δt. Linear stability, such as the A-stability, $A(\alpha)$-stability, or L-stability issues for the deferred correction methods were studied in, e.g. [18, 74, 111]. The SSP properties of deferred correction methods were first studied in [71]. In [57] it was shown that deferred correction methods are in fact Runge–Kutta methods, and thus all the order barriers and SSP bounds apply. In particular, the order barrier of four for explicit methods is unpleasant, as the major advantages of deferred correction methods are in the high order methods. For this reason, downwinding is desirable to extend the SSP properties of deferred correction methods to higher order methods.

The $(p+1)$-th order deferred correction time discretization of (2.2) can be formulated as follows. We first divide the time step $[t^n, t^{n+1}]$ where $t^{n+1} = t^n + \Delta t$ into s subintervals by choosing the points $t^{(m)}$ for $m = 0, 1, \cdots, s$ such that $t^n = t^{(0)} < t^{(1)} < \cdots < t^{(m)} < \cdots < t^{(s)} = t^{n+1}$. We use $\Delta t^{(m)} = t^{(m+1)} - t^{(m)}$ to denote the sub-time step and $u_k^{(m)}$ to denote the k-th order approximation to $u(t^{(m)})$. The nodes $t^{(m)}$ can be chosen equally spaced, or as the Chebyshev Gauss-Lobatto nodes on $[t^n, t^{n+1}]$ for high order accurate deferred correction schemes to avoid possible instability associated with interpolation on equally spaced points. Starting from u^n, the deferred correction algorithm to calculate u^{n+1} is in the following.

Compute the initial approximation
$$u_1^{(0)} = u^n.$$
Use the forward Euler method to compute a first order accurate approximate solution u_1 at the nodes $\{t^{(m)}\}_{m=1}^s$:
For $m = 0, \cdots, s-1$

$$u_1^{(m+1)} = u_1^{(m)} + \Delta t^{(m)} F(u_1^{(m)}). \tag{10.9}$$

Compute successive corrections

 For $k = 1, \cdots, s$
 $u_{k+1}^{(0)} = u^n$.
  For $m = 0, \cdots, s-1$

$$u_{k+1}^{(m+1)} = u_{k+1}^{(m)} + \theta_k \Delta t^{(m)} (F(u_{k+1}^{(m)}) - F(u_k^{(m)})) + I_m^{m+1}(F(u_k)), \tag{10.10}$$

where

$$0 \le \theta_k \le 1 \tag{10.11}$$

and $I_m^{m+1}(L(u_k))$ is the integral of the sth degree interpolating polynomial on the $s+1$ points $(t^{(\ell)}, L(u_k^{(\ell)}))_{\ell=0}^s$ over the subinterval $[t^{(m)}, t^{(m+1)}]$, which is the numerical quadrature approximation of

$$\int_{t^{(m)}}^{t^{(m+1)}} F(u(\tau))d\tau. \tag{10.12}$$

Finally we have $u^{n+1} = u_{s+1}^{(s)}$.

The scheme described above with $\theta_k = 1$ is the one discussed in [18, 74]. In [111], the scheme is also discussed with general $0 \le \theta_k \le 1$ to enhance linear stability. The term with the coefficient θ_k does not change the order of accuracy.

In [71], the SSP properties of the DC time discretization for the second, third and fourth order accuracy ($p = 1, 2, 3$), were studied. This was just a preliminary study, as the real advantage of the deferred correction time discretization is expected to show more clearly for much higher order of accuracy (the *spectral* DC method). The findings in [71] can be summarized below:

- The second order ($p = 1$) deferred correction time discretization has no subgrid point inside the interval $[t^n, t^{n+1}]$, and it is identical to the optimal second order Runge–Kutta SSP scheme SSPRK(2,2) given in Equation (2.16).
- For the third order ($p = 2$) deferred correction time discretization, there is only one subgrid point inside the interval $[t^n, t^{n+1}]$. By symmetry, this point should be placed in the middle, that is, $t^{(0)} = t^n$, $t^{(1)} = t^n + \frac{1}{2}\Delta t$, $t^{(2)} = t^{n+1}$.

A numerical optimization procedure can then be performed to search for the SSP scheme with the largest SSP coefficient. Unfortunately, it seems that negative β must appear hence the operator $\tilde{F}$ must be used. A SSP scheme with ten evaluations of F or $\tilde{F}$ is found to have a SSP coefficient $\tilde{C} = 1.2956$. Several other third order SSP deferred correction schemes are also found in [71] within specific subclasses, however none of them has an impressive SSP coefficient.

- For the fourth order ($p = 3$) deferred correction time discretization, there are two subgrid points inside the interval $[t^n, t^{n+1}]$. By symmetry, these two points should be placed at $t^{(1)} = t^n + a\Delta t$ and $t^{(2)} = t^n + (1-a)\Delta t$ respectively for $0 < a < \frac{1}{2}$. For example, the choice $a = \frac{5-\sqrt{5}}{10}$ would generate the standard Chebyshev Gauss-Lobatto nodes.

  As before, a numerical optimization procedure can then be performed to search for the SSP scheme with the largest SSP coefficient. Unfortunately, it again seems that negative β must appear hence the operator $\tilde{F}$ must be used. A SSP scheme with 17 evaluations of F or $\tilde{F}$ is found to have a SSP coefficient $\tilde{C} = 1.0319$. Several other fourth order SSP deferred correction schemes are also found in [71] within specific subclasses, however none of them has an impressive SSP coefficient.

It would seem from the results in [71] that low order deferred correction schemes are not competitive in terms of SSP properties when comparing with Runge–Kutta methods. However, this may not be the case with higher order deferred correction schemes.

We observe that the method (10.9)-(10.10) can be reinterpreted as a special class of Runge–Kutta method. It is easiest to write it in the Shu-Osher form. This involves nothing more than a change of notation, relabeling $u_{k+1}^{(m)}$ as $u^{(sk+m)}$. Comparison of the two forms reveals that the non-zero coefficients are

$$\left.\begin{array}{l} \alpha_{i,i-1} = 1 \\ \beta_{i,i-1} = \Delta t^{(i-1)}/\Delta t \end{array}\right\} \quad \text{for } 1 \le i \le s$$

$$\left.\begin{array}{l} \alpha_{sk+1,0} = 1 \\ \beta_{sk+1,s(k-1)+i} = C_i^0/\Delta t \end{array}\right\} \quad \text{for } \begin{cases} 1 \le k \le s \\ 0 \le i \le s \end{cases}$$

$$\left.\begin{array}{l} \alpha_{sk+m+1,sk+m} = 1 \\ \beta_{sk+m+1,s(k-1)} = C_i^m/\Delta t \end{array}\right\} \quad \text{for } \begin{cases} 1 \le m \le s-1 \\ 1 \le k \le s \\ 0 \le i \ne m \le s \end{cases}$$

$$\left.\begin{array}{l} \beta_{sk+m+1,sk+m} = \theta_k \Delta t^{(m)}/\Delta t \\ \beta_{sk+m+1,s(k-1)+m} = C_m^m/\Delta t - \theta_k \Delta t^{(m)}/\Delta t \end{array}\right\} \quad \text{for } \begin{cases} 1 \le k \le s \\ 1 \le m \le s-1 \end{cases}$$

where

$$I_m^{m+1}(L(u_k)) = \sum_{i=0}^{s} C_i^m u_k^{(i)}.$$

Using this, we can immediately obtain the Butcher array $(\mathbf{A}, \boldsymbol{b})$ and consider the much simpler problem of optimizing $\tilde{C}$ over the free parameters. For instance, for third order methods, this leads to a two-parameter optimization problem; the same problem was written (after some work) in terms of 16 free parameters in [71]. For fourth order methods the current approach leads to a six-parameter problem versus 69 parameters in [71].

The numerical optimization procedure described above can be applied to deferred correction schemes of any order to explore their SSP property. The algebra and computational cost for this procedure become very complicated, even for the fourth order methods considered in [71], if the traditional SSP theory is used. However, the analysis is relatively straightforward using the theory of absolutely monotonic methods.

Noting that spectral deferred correction methods can be written as explicit Runge–Kutta methods, we can immediately conclude that downwind operators will be required in order for explicit spectral deferred correction methods to be SSP if they are of order greater than four. Similarly, implicit spectral deferred correction methods cannot be SSP without downwinding if their order exceeds six. The use of downwind operators will allow us to extend the SSP properties of deferred correction methods to high order methods.

Chapter 11

Applications

In this chapter we will present some applications of the SSP time discretizations.

11.1 TVD schemes

The earliest application which motivated the design of the first SSP schemes is the total variation diminishing (TVD) schemes in solving hyperbolic conservation laws

$$u_t + f(u)_x = 0. \tag{11.1}$$

A finite difference, finite volume or discontinuous Galerkin finite element spatial discretization of (11.1) results in the following method of lines ODE system

$$\frac{d}{dt}u_j(t) = F(u)_j \tag{11.2}$$

where u_j for $j = 1, \cdots, N$ represent the approximations to the point values $u(x_j, t)$ for finite difference, or to the cell averages $\bar{u}_j(t) = \frac{1}{\Delta x} \int_{x_j - \Delta x/2}^{x_j + \Delta x/2} u(x, t)dx$ for finite volume or discontinuous Galerkin methods. The total variation semi-norm of the numerical solution u_j is defined as

$$TV(u) = \sum_j |u_{j+1} - u_j|. \tag{11.3}$$

If the solution u to the semi-discrete scheme (11.2) satisfies

$$\frac{d}{dt}TV(u) \leq 0, \tag{11.4}$$

then the scheme (11.2) is referred to as a TVD scheme. Since the exact solution of the PDE (11.1) has the total variation diminishing property (see, e.g. [94]), it is desirable to have the scheme (11.2) also share the same property.

It turns out that it is relatively easy to obtain not only the semi-discrete schemes (11.2) to satisfy the TVD condition (11.4), but also fully discretized scheme with the forward Euler time discretization

$$u_j^{n+1} = u_j^n + \Delta t F(u^n)_j \tag{11.5}$$

to satisfy similar TVD property

$$TV(u^{n+1}) \leq TV(u^n). \tag{11.6}$$

In fact, the following simple lemma, due to Harten [34], gives a general framework to construct TVD schemes.

Lemma 11.1. *(Harten). If $F(u)_j$ in the scheme (11.2) or (11.5) can be written in the form*

$$F(u)_j = C_{j+\frac{1}{2}}(u_{j+1} - u_j) - D_{j-\frac{1}{2}}(u_j - u_{j-1}) \tag{11.7}$$

with periodic or compactly supported boundary conditions, where $C_{j+\frac{1}{2}}$ and $D_{j-\frac{1}{2}}$ may be nonlinear functions of the grid values u_i for $i = j-p, \cdots, j+q$ with some $p, q \geq 0$, satisfying

$$C_{j+\frac{1}{2}} \geq 0, \qquad D_{j+\frac{1}{2}} \geq 0, \qquad \forall j, \tag{11.8}$$

then the semi-discrete scheme (11.2) is TVD. Furthermore, if we also have

$$\Delta t(C_{j+\frac{1}{2}} + D_{j+\frac{1}{2}}) \leq 1, \qquad \forall j, \tag{11.9}$$

then the fully discrete scheme (11.5) is TVD.

Proof. We only prove the result for the fully discrete scheme (11.5). Taking the forward difference operation on the scheme (11.5), considering (11.7), we have

$$\Delta_+ u_j^{n+1}$$

$$= \Delta_+ u_j^n + \Delta t \left(C_{j+\frac{3}{2}} \Delta_+ u_{j+1}^n - C_{j+\frac{1}{2}} \Delta_+ u_j^n - D_{j+\frac{1}{2}} \Delta_+ u_j^n + D_{j-\frac{1}{2}} \Delta_+ u_{j-1}^n \right)$$

$$= \left(1 - \Delta t(C_{j+\frac{1}{2}} + D_{j+\frac{1}{2}}) \right) \Delta_+ u_j^n + \Delta t C_{j+\frac{3}{2}} \Delta_+ u_{j+1}^n + \Delta t D_{j-\frac{1}{2}} \Delta_+ u_{j-1}^n$$

where the forward difference operator is defined by

$$\Delta_+ u_j = u_{j+1} - u_j.$$

Thanks to (11.8) and using the periodic or compactly supported boundary condition, we can take the absolute value on both sides of the above equality and sum up over j to obtain

$$\sum_j |\Delta_+ u_j^{n+1}| \le \sum_j \left(1 - \Delta t(C_{j+\frac{1}{2}} + D_{j+\frac{1}{2}})\right) |\Delta_+ u_j^n|$$

$$+ \sum_j \Delta t C_{j+\frac{1}{2}} |\Delta_+ u_j^n| + \sum_i \Delta t D_{j+\frac{1}{2}} |\Delta_+ u_j^n|$$

$$= \sum_j |\Delta_+ u_j^n|.$$

This finishes the proof. $\square$

Harten's lemma has led to many high order accurate finite difference, finite volume, discontinuous Galerkin, even spectral TVD schemes in the form of (11.2) or (11.5), when suitable nonlinear limiters are applied to the spatial discretization, see, e.g. [34, 76, 78, 13, 9]. Here, high order accuracy refers to the fact that

$$F(u)_j = -f(u)_x|_{x=x_j} + O(\Delta x^k)$$

for a large k when u is smooth and monotone. However, because a forward Euler time discretization is used in the fully discrete scheme (11.5), the temporal accuracy is still only first order. If a high order temporal accuracy is achieved via the Lax-Wendroff procedure [65, 82], it is very difficult to prove the TVD property of the resulting scheme, either by the Harten's lemma or directly.

Now with the SSP time discretization, we can easily obtain fully high order (in space and time) schemes which are TVD. Most of the high order in time TVD schemes in the literature have used SSP time discretizations. This is apparently the most successful application of SSP methods.

11.2   Maximum principle satisfying schemes and positivity preserving schemes

Even though the TVD schemes described in the previous section have been very successful in applications, they do have the following two restrictions:

(1) The entropy solutions to multi-dimensional conservation laws do satisfy the same TVD property. However, no TVD schemes can have higher than first order accuracy in multi-dimensions [24].

(2) Even in one space dimension, TVD schemes necessarily degenerate to at most first order accuracy at *smooth* extrema [77]. Therefore, even though very high order TVD schemes can be designed for smooth and monotone solutions (e.g. in [78]), for generic solutions which contain smooth extrema (for example all smooth solutions with periodic or compact supported boundary conditions must contain extrema), TVD schemes are at most second order accurate in the L^1 norm.

In order to overcome the first difficulty, TVD schemes are formally generalized to two or higher dimensions. For example, in two dimensions the semi-discrete scheme (11.2) becomes

$$\frac{d}{dt}u_{ij}(t) = F(u)_{ij}, \tag{11.10}$$

with its forward Euler version

$$u_{ij}^{n+1} = u_{ij}^n + \Delta t F(u^n)_{ij}, \tag{11.11}$$

where u_{ij} for $i = 1, \cdots, N_x$ and $j = 1, \cdots, N_y$ now represent the approximations to the point values $u(x_i, y_j, t)$ for finite difference, or to the cell averages $\bar{u}_{ij}(t) = \frac{1}{\Delta x \Delta y} \int_{y_j - \Delta y/2}^{y_j + \Delta y/2} \int_{x_i - \Delta x/2}^{x_i + \Delta x/2} u(x, y, t) dx dy$ for finite volume or discontinuous Galerkin methods. The Harten's form of the spatial operator (11.7) is now generalized to

$$F(u)_{ij} = C_{i+\frac{1}{2},j}(u_{i+1,j} - u_{ij}) - D_{i-\frac{1}{2},j}(u_{ij} - u_{i-1,j})$$
$$+ C_{i,j+\frac{1}{2}}(u_{i,j+1} - u_{ij}) - D_{i,j-\frac{1}{2}}(u_{ij} - u_{i,j-1}), \tag{11.12}$$

still with the condition

$$C_{i+\frac{1}{2},j} \geq 0, \quad D_{i-\frac{1}{2},j} \geq 0, \quad C_{i,j+\frac{1}{2}} \geq 0, \quad D_{i,j-\frac{1}{2}} \geq 0, \quad \forall j. \tag{11.13}$$

The time step restriction (11.9) is generalized to

$$\Delta t(C_{i+\frac{1}{2},j} + D_{i-\frac{1}{2},j} + C_{i,j+\frac{1}{2}} + D_{i,j-\frac{1}{2}}) \leq 1, \quad \forall j. \tag{11.14}$$

Even though the strict TVD property can no longer hold because of the Goodman-LeVeque theorem [24], maximum-norm stability can still be easily obtained through the following modified Harten's lemma.

Lemma 11.2. *(Generalized Harten's lemma). If $F(u)_{ij}$ in the scheme (11.10) or (11.11) can be written in the form (11.12), with periodic or compactly supported boundary conditions, where $C_{i+\frac{1}{2},j}$, $D_{i-\frac{1}{2},j}$, $C_{i,j+\frac{1}{2}}$ and $D_{i,j-\frac{1}{2}}$ satisfy (11.13), then the semi-discrete scheme (11.10) satisfies the maximum principle*

$$\max_{i,j} u_{ij}(t) \leq \max_{i,j} u_{ij}(0), \qquad \min_{i,j} u_{ij}(t) \geq \min_{i,j} u_{ij}(0), \qquad t > 0.$$

Furthermore, if the time step restriction (11.14) is satisfied, then the fully discrete scheme (11.11) satisfies the maximum principle

$$\max_{i,j} u_{ij}^{n+1} \leq \max_{i,j} u_{ij}^n, \qquad \min_{i,j} u_{ij}^{n+1} \geq \min_{i,j} u_{ij}^n, \qquad n \geq 0. \qquad (11.15)$$

The proof of this lemma is similar to that for the original Harten's lemma.

Using this lemma and SSP time discretizations, high order accurate multi-dimensional schemes which satisfy such maximum norm stability have been developed in the literature, for example the discontinuous Galerkin method in [12].

However, even though the schemes described above satisfy the maximum principle and can be arbitrarily high order accurate in smooth and monotone regions, at smooth extrema they still degenerate to at most second order accuracy. There is a very simple proof of this fact, due to Ami Harten (the third author, Chi-Wang Shu, learned this from a class at UCLA taught by Harten in the mid 1980s): Suppose we are solving the one dimensional linear conservation law

$$u_t + u_x = 0, \qquad u(x,0) = \sin(x) \qquad (11.16)$$

with periodic boundary conditions. Let us choose a uniform mesh with $\Delta x = \frac{\pi}{2N-1}$ and $x_j = j\Delta x$. Clearly, for the initial condition at the grid points,

$$u_j^0 = \sin(x_j) \leq \sin(x_N) = \sin\left(\frac{\pi}{2} + \frac{\Delta x}{2}\right) \leq 1 - \alpha \Delta x^2, \qquad \forall j$$

with a constant $\alpha > 0$. If our scheme satisfies the maximum principle

$$\max_j u_j^{n+1} \leq \max_j u_j^n, \qquad (11.17)$$

then we have

$$\max_j u_j^1 \leq \max_j u_j^0 \leq 1 - \alpha \Delta x^2.$$

However, if we take the time step $\Delta t = \frac{1}{2}\Delta x$, the exact solution at the Nth point at the first time step is

$$u(x_N, \Delta t) = \sin(x_N + \Delta t) = \sin\left(\frac{\pi}{2}\right) = 1.$$

Therefore the error at the Nth point is at least

$$|u(x_N, \Delta t) - u_N^1| = u(x_N, \Delta t) - u_N^1 \geq \alpha \Delta x^2.$$

That is, even after one step, the error at the smooth extrema is already at most second order. Note that this proof uses only the definition of the maximum principle (11.17) and is independent of the actual scheme. Therefore, any paper claiming to have uniform high order (larger than second order) accurate schemes which satisfy the maximum principle (11.17) must be incorrect! Similar example can also be constructed for finite volume schemes where u_j^n approximate the cell averages.

This simple example indicates that we cannot have uniformly high order schemes which satisfy the maximum principle in the sense of (11.15). In [113, 114], following the idea of [90], the maximum principle is defined alternatively as

$$\max_x u^{n+1}(x) \leq \max_x u^n(x), \qquad \min_x u^{n+1}(x) \geq \min_x u^n(x), \qquad n \geq 0$$
(11.18)

where $u^n(x)$ is the piecewise polynomial either reconstructed in a finite volume scheme or evolved in a discontinuous Galerkin scheme. We can then construct uniformly accurate (in space) schemes whose forward Euler time discretization satisfies such maximum principle. SSP time discretization can then be applied to enhance the temporal accuracy while maintaining the same maximum principle. We describe this procedure briefly below.

First, it is well known that a first order monotone scheme

$$u_j^{n+1} = H_\lambda(u_{j-1}^n, u_j^n, u_{j+1}^n) = u_j^n - \lambda\left(\hat{f}(u_j^n, u_{j+1}^n) - \hat{f}(u_{j-1}^n, u_j^n)\right), \quad (11.19)$$

where $\lambda = \frac{\Delta t}{\Delta x}$, and $\hat{f}(a,b)$ is a monotone flux, namely it is Lipschitz continuous in both argument, non-decreasing in the first argument and non-increasing in the second argument, and consistent $\hat{f}(a,a) = f(a)$, satisfies the maximum principle

$$\min_i u_i^n \leq u_j^{n+1} \leq \max_i u_i^n$$

under the CFL condition

$$(|\hat{f}_1| + |\hat{f}_2|)\lambda \leq 1 \qquad (11.20)$$

where $|\hat{f}_1|$ and $|\hat{f}_2|$ are the Lipschitz constants of the numerical flux $\hat{f}$ with respect to the first and second arguments respectively. This can be easily verified by noticing that the function H_λ in the scheme (11.19) is a monotonically increasing (non-decreasing) function of all its three arguments under the CFL condition (11.20).

We now move to higher order methods. A finite volume scheme, or the scheme satisfied by the cell averages of a discontinuous Galerkin method, takes the form

$$\bar{u}_j^{n+1} = \bar{u}_j^n - \lambda\left(\hat{f}(u_{j+\frac{1}{2}}^-, u_{j+\frac{1}{2}}^+) - \hat{f}(u_{j-\frac{1}{2}}^-, u_{j-\frac{1}{2}}^+)\right), \qquad (11.21)$$

where we assume that there is a piecewise polynomial $u^n(x)$, either reconstructed in a finite volume method or evolved in a discontinuous Galerkin method, which is a polynomial $p_j(x)$ of degree k in the cell $I_j = (x_{j-\frac{1}{2}}, x_{j+\frac{1}{2}})$, such that

$$\bar{u}_j^n = \frac{1}{\Delta x} \int_{x_{j-\frac{1}{2}}}^{x_{j+\frac{1}{2}}} p_j(x)dx, \qquad u_{j-\frac{1}{2}}^+ = p_j(x_{j-\frac{1}{2}}), \qquad u_{j+\frac{1}{2}}^- = p_j(x_{j+\frac{1}{2}}).$$

We take a q-point Legendre Gauss-Lobatto quadrature rule which is exact for polynomials of degree k, then

$$\bar{u}_j^n = \sum_{\ell=0}^{q} \omega_\ell p_j(y_\ell)$$

where y_ℓ are the Legendre Gauss-Lobatto quadrature points, with $y_0 = x_{j-\frac{1}{2}}$, $y_q = x_{j+\frac{1}{2}}$, and ω_ℓ are the quadrature weights which are all positive.

In [113, 114], a simple scaling limiter is designed to maintain the point values $p_j(y_\ell)$ between the lower bound m and upper bound M, provided (i) the cell averages $\bar{u}_j^n$ are within these bounds; (ii) the cell averages $\bar{u}_j^n$ are high order approximations to the cell averages of the exact solution $u(x, t^n)$; and (iii) the exact solution $u(x, t^n)$ is within these bounds for all x. This limiter is proved in [113, 114] to maintain uniform high order accuracy, and it is very easy to implement.

Now, the scheme (11.21) can be rewritten as

$$\bar{u}_j^{n+1} = \omega_q \left[u_{j+\frac{1}{2}}^- - \frac{\lambda}{\omega_q} \left(\hat{f}(u_{j+\frac{1}{2}}^-, u_{j+\frac{1}{2}}^+) - \hat{f}(u_{j-\frac{1}{2}}^+, u_{j+\frac{1}{2}}^-) \right) \right]$$

$$+ \omega_0 \left[u_{j-\frac{1}{2}}^+ - \frac{\lambda}{\omega_0} \left(\hat{f}(u_{j-\frac{1}{2}}^+, u_{j+\frac{1}{2}}^-) - \hat{f}(u_{j-\frac{1}{2}}^-, u_{j-\frac{1}{2}}^+) \right) \right]$$

$$+ \sum_{\ell=1}^{q-1} \omega_\ell p_j(y_\ell)$$

$$= H_{\lambda/\omega_q}(u_{j-\frac{1}{2}}^+, u_{j+\frac{1}{2}}^-, u_{j+\frac{1}{2}}^+) + H_{\lambda/\omega_0}(u_{j-\frac{1}{2}}^-, u_{j-\frac{1}{2}}^+, u_{j+\frac{1}{2}}^-) + \sum_{\ell=1}^{q-1} \omega_\ell p_j(y_\ell).$$

Clearly, by the monotonicity of the function H, all the terms on the right-hand side of the last equality are within the bound m and M, provided that a reduced CFL condition

$$(|\hat{f}_1| + |\hat{f}_2|)\lambda \leq \omega_0$$

is satisfied (note that $\omega_0 = \omega_q$). Therefore, $\bar{u}_j^{n+1}$ is also within the bound m and M as it is a convex combination of these terms with $\sum_{\ell=0}^{q} \omega_\ell = 1$. Now, the limiter described above can be applied again to bring the new piecewise polynomial at time level $n + 1$ to stay within the bound m and M at the Legendre Gauss-Lobatto quadrature points while maintaining uniform high order accuracy, thus finishing a complete cycle of the scheme design.

We have thus obtained a uniformly high order (in space) scheme with forward Euler time discretization, which satisfies the maximum principle. An application of any SSP time discretization then upgrades the temporal accuracy also to high order while maintaining the same maximum principle. This idea has been extended to two space dimensions [114], incompressible flows in vorticity-streamfunction formulation and any passive convection in an incompressible velocity field [114], positivity preserving (for density and pressure) schemes for Euler equations of compressible gas dynamics [115], positivity preserving (for water height) schemes for shallow water equations [112], and maximum principle satisfying and positivity preserving schemes on unstructured meshes [116].

11.3 Coercive approximations

An application of SSP methods to coercive approximations is discussed in [27]. We consider the linear system of ODEs of the general form, with possibly variable, time-dependent coefficients,

$$\frac{d}{dt}u(t) = L(t)u(t). \tag{11.22}$$

As an example we refer to [25], where the far-from-normal character of the spectral differentiation matrices defies the straightforward von Neumann stability analysis when augmented with high order time discretizations.

We begin our stability study with the first order forward Euler scheme

$$u^{n+1} = u^n + \Delta t L(t^n)u^n. \tag{11.23}$$

Of course, we could also have variable time steps Δt_j such that $t^n := \sum_{j=0}^{n-1} \Delta t_j$. In the following we will use $\langle \cdot, \cdot \rangle$ to denote the usual Euclidean inner product. Taking L^2 norms on both sides of (11.23) one finds

$$|u^{n+1}|^2 = |u^n|^2 + 2\Delta t Re\langle L(t^n)u^n, u^n \rangle + (\Delta t)^2 |L(t^n)u^n|^2,$$

and hence strong stability holds,

$$|u^{n+1}| \leq |u^n|, \tag{11.24}$$

provided the following restriction on the time step Δt is met,

$$\Delta t \leq -2Re\langle L(t^n)u^n, u^n \rangle / |L(t^n)u^n|^2.$$

Following Levy and Tadmor [69] we make the following assumption.

Assumption 11.1. (Coercivity). The operator $L(t)$ is (uniformly) *coercive* in the sense that there exists $\eta(t) > 0$ such that

$$\eta(t) := \inf_{|u|=1} -\frac{Re\langle L(t)u, u\rangle}{|L(t)u|^2} > 0. \tag{11.25}$$

We conclude that for coercive L's, the forward Euler scheme is strongly stable, i.e. (11.24) holds for the solution u^n to (11.23), if

$$\Delta t \leq 2\eta(t^n). \tag{11.26}$$

In a generic case, $L(t^n)$ represents a spatial operator with a coercivity-bound $\eta(t^n)$, which is proportional to some power of the smallest spatial scale. In this context the above restriction on the time step amounts to the usual CFL stability condition.

We observe that the coercivity constant, η, is an upper bound in the size of L; indeed, by Cauchy-Schwartz, $\eta(t) \leq |L(t)u| \cdot |u|/|L(t)u|^2$ and hence

$$\|L(t)\| = \sup_u \frac{|L(t)u|}{|u|} \leq \frac{1}{\eta(t)}. \tag{11.27}$$

Now the SSP theory trivially implies that a high order SSP Runge–Kutta method for linear problems, with SSP coefficient (or, more accurately, linear threshold value) $R = 1$ (see Chapter 4), would provide a strongly stable solution satisfying (11.24) under the same time step restriction (11.26), if L does not depend on t.

For time dependent L, starting with second order and higher the Runge–Kutta intermediate steps depend on the time variation of $L(\cdot)$, and hence we require a minimal smoothness in time. We therefore make the following assumption.

Assumption 11.2. (Lipschitz regularity). We assume that $L(\cdot)$ is Lipschitz. Thus, there exists a constant $K > 0$ such that

$$\|L(t) - L(s)\| \leq \frac{K}{\eta(t)}|t - s|. \tag{11.28}$$

Under this assumption, we can easily prove the following result, see [27] for the detailed proof.

Proposition 11.1. *Consider the coercive systems of ODEs, (11.22)-(11.25), with Lipschitz continuous coefficients (11.28). Then the numerical* *solution obtained with the classical fourth order Runge–Kutta method is stable under the CFL restriction condition (11.26) and the following estimate holds*

$$|u^n| \leq e^{3Kt_n}|u^0|. \tag{11.29}$$

In [69], a result along these lines was introduced by Levy and Tadmor for constant coefficient linear operators $L(t) = L$ in (11.22). They demonstrated the strong stability of the constant coefficients p order Runge–Kutta scheme under CFL condition $\Delta t_n \leq K_p \eta(t^n)$ with $K_4 = 1/31$ (in Theorem 3.3 of [69]). Using the SSP theory for linear constant coefficient operators we can improve upon those results in both simplicity and generality.

Proposition 11.2. *Consider the coercive systems of ODEs, (11.22)-(11.25), with a constant coefficient operator $L(t) = L$. Then the numerical solution obtained with any of the p-stage, pth order SSP Runge–Kutta schemes in Chapter 4 is stable under the CFL restriction condition*

$$\Delta t_n \leq K_p \eta(t^n)$$

with a uniform $K_p = 2$.

Note that whenever Equation (11.22) can be manipulated into, or approximated by, a linear constant coefficient system which satisfies the forward Euler condition (11.23), then Proposition 11.2 can be used.

Remark 11.1. Although SSP methods give us improved results over [69], significantly stronger results can be obtained in this case by using the theory of monotonicity preservation in inner product norms (see [36]). For instance, strict monotonicity can be shown to hold, in Proposition 11.1, for the same methods under the same time step restriction (11.26), even for nonlinear, time dependent systems, without the Lipschitz assumption (11.28).

In general, the theory detailed in [36] provides a better alternative to SSP theory when the norm in question is generated by an inner product. This theory should be used whenever dealing with an inner product norm, even for a linear constant coefficient system, as it allows larger time steps under more relaxed assumptions (see Table 1 of [36]).

Bibliography

[1] P. Albrecht. The Runge–Kutta theory in a nutshell. *SIAM Journal on Numerical Analysis*, 33:1712–1735, 1996.

[2] D. Balsara and C.-W. Shu. Monotonicity preserving weighted essentially non-oscillatory schemes with increasingly high order of accuracy. *Journal of Computational Physics*, 160:405–452, 2000.

[3] A. Bellen and Z. Jackiewicz and M. Zennaro. Contractivity of waveform relaxation Runge–Kutta iterations and related limit methods for dissipative systems in the maximum norm. *SIAM Journal on Numerical Analysis*, 31:499–523, 1994.

[4] A. Bellen and L. Torelli. Unconditional contractivity in the maximum norm of diagonally split Runge-Kutta methods. *SIAM Journal on Numerical Analysis*, 34:528–543, 1997

[5] C. Bolley and M. Crouzeix. Conservation de la positivite lors de la discretisation des problemes d'evolution paraboliques. *R.A.I.R.O. Anal. Numer.* 12:237–245, 1978.

[6] J.C. Butcher. On the implementation of implicit Runge-Kutta methods. *BIT*, 6:237–240, 1976.

[7] J. C. Butcher and S. Tracogna. Order conditions for two-step Runge–Kutta methods. *Applied Numerical Mathematics*, 24:351–364, 1997.

[8] J.C. Butcher. *Numerical Methods for Ordinary Differential Equations*. Wiley, 2003.

[9] W. Cai, D. Gottlieb and C.-W. Shu. Essentially nonoscillatory spectral Fourier methods for shock wave calculations. *Mathematics of Computation*, 52:389–410, 1989.

[10] M. Calvo, J.M. Franco, and L. Rández. A new minimum storage Runge-Kutta scheme for computational acoustics. *Journal of Computational Physics*, 201:1–12, 2004.

[11] M.-H. Chen, B. Cockburn and F. Reitich. High-order RKDG methods for computational electromagnetics. *Journal of Scientific Computing*, 22-23:205–226, 2005.

[12] B. Cockburn, S. Hou and C.-W. Shu. The Runge-Kutta local projection discontinuous Galerkin finite element method for conservation laws IV: the

multidimensional case. *Mathematics of Computation*, 54:545–581, 1990.

[13] B. Cockburn and C.-W. Shu. TVB Runge-Kutta local projection discontinuous Galerkin finite element method for conservation laws II: general framework. *Mathematics of Computation*, 52:411–435, 1989.

[14] B. Cockburn and C.-W. Shu. Runge-Kutta Discontinuous Galerkin methods for convection-dominated problems. *Journal of Scientific Computing*, 16:173–261, 2001.

[15] E. Constantinescu and A. Sandu. Optimal explicit strong-stability-preserving general linear methods. *SIAM Journal on Scientific Computing*, 32(5):3130–3150, 2010.

[16] G. Dahlquist and R. Jeltsch. Generalized disks of contractivity for explicit and implicit Runge-Kutta methods. Technical report, Department of Numerical Analysis and Computational Science, Royal Institute of Technology, Stockholm, 1979.

[17] K. Dekker and J.G. Verwer. *Stability of Runge-Kutta methods for stiff nonlinear differential equations*, Volume 2 of *CWI Monographs*. North-Holland Publishing Co., Amsterdam, 1984.

[18] A. Dutt, L. Greengard and V. Rokhlin. Spectral deferred correction methods for ordinary differential equations. *BIT*, 40:241–266, 2000.

[19] L. Ferracina and M.N. Spijker. Stepsize restrictions for the total-variation-diminishing property in general Runge-Kutta methods. *SIAM Journal of Numerical Analysis*, 42:1073–1093, 2004.

[20] L. Ferracina and M.N. Spijker. Computing optimal monotonicity-preserving Runge-Kutta methods. Technical Report MI2005-07, Mathematical Institute, Leiden University, 2005.

[21] L. Ferracina and M.N. Spijker. An extension and analysis of the Shu-Osher representation of Runge-Kutta methods. *Mathematics of Computation*, 249:201–219, 2005.

[22] L. Ferracina and M.N. Spijker. Strong stability of singly-diagonally-implicit Runge-Kutta methods. *Applied Numerical Mathematics*, 2008. doi: 10.1016/j.apnum.2007.10.004

[23] S. Gill. A process for the step-by-step integration of differential equations in an automatic digital computing machine. *Proceedings of the Cambridge Philosophical Society*, 47:96–108, 1950.

[24] J.B. Goodman and R.J. LeVeque. On the accuracy of stable schemes for 2D scalar conservation laws. *Mathematics of Computation*, 45:15–21, 1985.

[25] D. Gottlieb and E. Tadmor. The CFL condition for spectral approximations to hyperbolic initial-boundary value problems. *Mathematics of Computation*, 56:565–588, 1991.

[26] S. Gottlieb and C.-W. Shu. Total variation diminishing Runge-Kutta schemes. *Mathematics of Computation*, 67:73–85, 1998.

[27] S. Gottlieb, C.-W. Shu and E. Tadmor. Strong stability preserving high-order time discretization methods. *SIAM Review*, 43:89–112, 2001.

[28] S. Gottlieb and L.J. Gottlieb. Strong stability preserving properties of Runge-Kutta time discretization methods for linear constant coefficient operators. *Journal of Scientific Computing*, 18:83–109, 2003.

[29] S. Gottlieb. On high order strong stability preserving Runge-Kutta and multi step time discretizations. *Journal of Scientific Computing*, 25:105–127, 2005.

[30] S. Gottlieb and S.J. Ruuth. Optimal strong-stability-preserving time-stepping schemes with fast downwind spatial discretizations. *Journal of Scientific Computing*, 27:289–303, 2006.

[31] S. Gottlieb, D.I. Ketcheson and C.-W. Shu. High order strong stability preserving time discretizations. *Journal of Scientific Computing*, 38:251–289, 2009.

[32] E. Hairer and G. Wanner. *Solving ordinary differential equations II: Stiff and differential-algebraic problems*, Volume 14 of *Springer Series in Computational Mathematics*. Springer-Verlag, Berlin, 1991.

[33] E. Hairer and G. Wanner. Order conditions for general two-step Runge-Kutta methods. *SIAM Journal on Numerical Analysis*, 34(6):2087–2089, 1997.

[34] A. Harten. High resolution schemes for hyperbolic conservation laws. *Journal of Computational Physics*, 49:357–393, 1983.

[35] I. Higueras. On strong stability preserving time discretization methods. *Journal of Scientific Computing*, 21:193–223, 2004.

[36] I. Higueras. Monotonicity for Runge-Kutta methods: inner-product norms. *Journal of Scientific Computing*, 24:97–117, 2005.

[37] I. Higueras. Representations of Runge-Kutta methods and strong stability preserving methods. *SIAM Journal on Numerical Analysis*, 43:924–948, 2005.

[38] I. Higueras. Strong stability for additive Runge-Kutta methods. *SIAM Journal on Numerical Analysis*, 44:1735–1758, 2006.

[39] I. Higueras. Characterizing Strong Stability Preserving Additive Runge-Kutta Methods. Journal of Scientific Computing, 39:115–128, 2009.

[40] Z. Horvath. Positivity of Runge-Kutta and Diagonally Split Runge-Kutta Methods Applied Numerical Mathematics, 28:309-326, 1998.

[41] Z. Horvath. On the positivity of matrix-vector products. *Linear Algebra and Its Applications*, 393:253–258, 2004.

[42] Z. Horvath. On the positivity step size threshold of Runge-Kutta methods. *Applied Numerical Mathematics*, 53:341–356, 2005.

[43] K.J.I. Hout. A note on unconditional maximum norm contractivity of diagonally split Runge-Kutta methods. *SIAM Journal on Numerical Analysis*, 33:1125–1134, 1996.

[44] C. Hu and C.-W. Shu. Weighted essentially non-oscillatory schemes on triangular meshes. *Journal of Computational Physics*, 150:97–127, 1999.

[45] F.Q. Hu, M.Y. Hussaini, and J.L. Manthey. Low-dissipation and low-dispersion Runge-Kutta schemes for computational acoustics. *Journal of Computational Physics*, 124:177–191, 1996.

[46] C. Huang. Strong stability preserving hybrid methods. *Applied Numerical Mathematics*, doi: 10.1016/j.apnum.2008.03.030, 2008.

[47] C. Huang. Strong stability preserving hybrid methods. *Applied Numerical Mathematics*, 59(5):891–904, 2009.

[48] W. Hundsdorfer and S.J. Ruuth. On monotonicity and boundedness proper-

ties of linear multistep methods. *Mathematics of Computation*, 75(254):655–672, 2005.

[49] W. Hundsdorfer, S.J. Ruuth and R.J. Spiteri. Monotonicity-preserving linear multistep methods. *SIAM Journal on Numerical Analysis*, 41:605–623, 2003.

[50] Z. Jackiewicz and S. Tracogna. A general class of two-step Runge-Kutta methods for ordinary differential equations. *SIAM Journal of Numerical Analysis*, 32:1390–1427, 1995.

[51] G.-S. Jiang and C.-W. Shu. Efficient implementation of weighted ENO schemes. *Journal of Computational Physics*, 126:202–228, 1996.

[52] C.A. Kennedy, M.H. Carpenter and R.M. Lewis. Low-storage, explicit Runge-Kutta schemes for the compressible Navier-Stokes equations. *Applied Numerical Mathematics*, 35:177–219, 2000.

[53] D.I. Ketcheson and A.C. Robinson. On the practical importance of the SSP property for Runge-Kutta time integrators for some common Godunov-type schemes. *International Journal for Numerical Methods in Fluids*, 48:271–303, 2005.

[54] D.I. Ketcheson, C.B. Macdonald and S. Gottlieb. Optimal implicit strong stability preserving Runge-Kutta methods. *Applied Numerical Mathematics*, 52(2): 373–392, 2009.

[55] D. I. Ketcheson. Highly efficient strong stability preserving Runge-Kutta methods with low-storage implementations. *SIAM Journal on Scientific Computing*, 30(4):2113–2136, 2008.

[56] D.I. Ketcheson. Computation of optimal monotonicity preserving general linear methods. *Mathematics of Computation*, 78: 1497–1513, 2009.

[57] D.I. Ketcheson. High order strong stability preserving time integrators and numerical wave propagation methods for hyperbolic PDEs. Ph.D. Thesis, University of Washington, 2009.

[58] D.I. Ketcheson, S. Gottlieb, and C.B. Macdonald. Strong stability preserving two-step Runge-Kutta methods. Submitted manuscript.

[59] D. I. Ketcheson. Runge-Kutta methods with minimum storage implementations. *Journal of Computational Physics*, 229(5):1763–1773, 2010.

[60] J. F. B. M. Kraaijevanger. Absolute monotonicity of polynomials occurring in the numerical solution of initial value problems. *Numerische Mathematik*, 48:303–322, 1986.

[61] J. F. B. M. Kraaijevanger and M. N. Spijker. Algebraic stability and error propagation in Runge-Kutta methods. *Applied Numerical Mathematics*, 5:71–87, 1989.

[62] J. F. B. M. Kraaijevanger. Contractivity of Runge-Kutta methods. *BIT*, 31:482–528, 1991.

[63] A. Kurganov and E. Tadmor. New high-resolution schemes for nonlinear conservation laws and convection-diffusion equations. *Journal of Computational Physics*, 160:241–282, 2000.

[64] J.D. Lambert, *Computational Methods in Ordinary Differential Equations*, John Wiley, New York, 1973.

[65] P.D. Lax and B. Wendroff. Systems of conservation laws. *Communications in Pure and Applied Mathematics*, 13:217–237, 1960.

[66] H.W.J. Lenferink. Contractivity-preserving explicit linear multistep methods. *Numerische Mathematik*, 55:213–223, 1989.

[67] H.W.J. Lenferink. Contractivity-preserving implicit linear multistep methods. *Mathematics of Computation*, 56:177–199, 1991.

[68] R.J. LeVeque, Numerical Methods for Conservation Laws, Birkhäuser Verlag, Basel, 1990.

[69] D. Levy and E. Tadmor. From semi-discrete to fully discrete: stability of Runge-Kutta schemes by the energy method. *SIAM Review*, 40:40–73, 1998.

[70] X.-D. Liu, S. Osher and T. Chan. Weighted essentially non-oscillatory schemes. *Journal of Computational Physics*, 115(1):200–212, 1994.

[71] Y. Liu, C.-W. Shu and M. Zhang. Strong stability preserving property of the deferred correction time discretization. *Journal of Computational Mathematics*. To appear.

[72] C.B. Macdonald. Constructing high-order Runge-Kutta methods with embedded strong-stability-preserving pairs. Master's thesis, Simon Fraser University, August 2003.

[73] C.B. Macdonald, S. Gottlieb, and S. Ruuth. A numerical study of diagonally split Runge-Kutta methods for PDEs with discontinuities. *Journal of Scientific Computing*, 36 (1): 89-112, 2008.

[74] M.L. Minion. Semi-implicit spectral deferred correction methods for ordinary differential equations. *SIAM Journal on Numerical Analysis*, 41:605–623, 2003.

[75] H. Nessyahu and E. Tadmor. Non-oscillatory central differencing for hyperbolic conservation laws. *Journal of Computational Physics*, 87:408–463, 1990.

[76] S. Osher. Convergence of generalized MUSCL schemes. *SIAM Journal on Numerical Analysis*, 22:947–961, 1985.

[77] S. Osher and S. Chakravarthy. High resolution schemes and the entropy condition. *SIAM Journal on Numerical Analysis*, 21:955–984, 1984.

[78] S. Osher and S. Chakravarthy. Very high order accurate TVD schemes. in *IMA Volumes in Mathematics and Its Applications*, 2:229–274, 1986, Springer-Verlag.

[79] S. Osher and J. Sethian. Fronts propagating with curvature dependent speed: algorithms based on Hamilton-Jacobi formulations. *Journal of Computational Physics*, 79:12–49, 1988.

[80] S. Osher and C.-W. Shu. High-order essentially nonoscillatory schemes for Hamilton-Jacobi equations. *SIAM Journal on Numerical Analysis*, 28:907–922, 1991.

[81] S. Osher and E. Tadmor. On the convergence of difference approximations to scalar conservation laws. *Mathematics of Computation*, 50:19–51, 1988.

[82] J. Qiu and C.-W. Shu. Finite difference WENO schemes with Lax-Wendroff-type time discretizations. *SIAM Journal on Scientific Computing*, 24:2185–2198, 2003.

[83] A. Ralston, A First Course in Numerical Analysis, McGraw-Hill, New York, 1965.

[84] S.J. Ruuth. Global optimization of explicit strong-stability-preserving

Runge-Kutta methods. *Mathematics of Computation*, 75:183–207, 2006.

[85] S.J. Ruuth and W. Hundsdorfer. High-order linear multistep methods with general monotonicity and boundedness properties. *Journal of Computational Physics*, 209:226–248, 2005.

[86] S.J. Ruuth and R.J. Spiteri. Two barriers on strong-stability-preserving time discretization methods. *Journal of Scientific Computation*, 17:211–220, 2002.

[87] S.J. Ruuth and R.J. Spiteri. High-order strong-stability-preserving Runge-Kutta methods with downwind-biased spatial discretizations. *SIAM Journal on Numerical Analysis*, 42:974–996, 2004.

[88] N.V. Sahinidis and M. Tawarmalani. BARON 7.2: Global Optimization of Mixed-Integer Nonlinear Programs. Users Manual (2004). Available at http://www.gams.com/dd/docs/solvers/baron.pdf

[89] J. Sand. Circle contractive linear multistep methods. *BIT* 26:114–122, 1986.

[90] R. Sanders. A third-order accurate variation nonexpansive difference scheme for single nonlinear conservation law. *Mathematics of Computation*, 51:535–558, 1988.

[91] C.-W. Shu. Total-variation diminishing time discretizations. *SIAM Journal on Scientific and Statistical Computing*, 9:1073–1084, 1988.

[92] C.-W. Shu and S. Osher. Efficient implementation of essentially non-oscillatory shock-capturing schemes. *Journal of Computational Physics*, 77:439–471, 1988.

[93] C.-W. Shu and S. Osher. Efficient implementation of essentially non-oscillatory shock capturing schemes II. *Journal of Computational Physics*, 83:32–78, 1989.

[94] J. Smoller, Shock Waves and Reaction-Diffusion Equations. Springer-Verlag, New York, 1994.

[95] M.N. Spijker. Contractivity in the numerical solution of initial value problems. *Numerische Mathematik*, 42:271–290, 1983.

[96] M.N. Spijker. Stepsize conditions for general monotonicity in numerical initial value problems. *SIAM Journal on Numerical Analysis*, 45:1226–1245, 2007.

[97] R. J. Spiteri and S. J. Ruuth. A new class of optimal high-order strong-stability-preserving time discretization methods. *SIAM Journal of Numerical Analysis*, 40:469–491, 2002.

[98] R. J. Spiteri and S. J. Ruuth. Nonlinear evolution using optimal fourth-order strong-stability-preserving Runge-Kutta methods. *Mathematics and Computers in Simulation*, 62:125–135, 2003.

[99] G. Strang. Accurate partial difference methods ii: nonlinear problems. *Numerische Mathematik*, 6:37–46, 1964.

[100] J.C. Strikwerda. *Finite Difference Schemes and Partial Differential Equations*. Cole Mathematics Series. Wadsworth and Brooks, California, 1989.

[101] P.K. Sweby. High resolution schemes using flux limiters for hyperbolic conservation laws. *SIAM Journal on Numerical Analysis*, 21:995–1011, 1984.

[102] E. Tadmor. *"Advanced Numerical Approximation of Nonlinear Hyperbolic Equations,"* Lectures Notes from CIME Course Cetraro, Italy, 1997, chapter on Approximate solutions of nonlinear conservation laws, pages 1–150.

Number 1697 in Lecture Notes in Mathematics. Springer-Verlag, 1998.

[103] K. Tselios and T.E. Simos. Optimized Runge-Kutta methods with minimal dispersion and dissipation for problems arising from computational acoustics. *Physics Letters A*, 363:38–47, 2007.

[104] J.A. van de Griend and J.F.B.M. Kraaijevanger. Absolute monotonicity of rational functions occurring in the numerical solution of initial value problems. *Numerische Mathematik*, 49:413–424, 1986.

[105] P.J. van der Houwen. Explicit Runge-Kutta formulas with increased stability boundaries. *Numerische Mathematik*, 20:149–164, 1972.

[106] B. van Leer. Towards the ultimate conservative difference scheme V. A second order sequel to Godunov's method. *Journal of Computational Physics*, 32:101–136, 1979.

[107] J. H. Verner. High-order explicit RungeKutta pairs with low stage order. *Applied Numerical Methods*, 22, 345357, 1996.

[108] J. H. Verner. Improved starting methods for two-step Runge–Kutta methods of stage-order p-3. *Applied Numerical Mathematics*, 56(3-4):388–396, 2006.

[109] J. H. Verner. Starting methods for two-step Runge–Kutta methods of stage-order 3 and order 6. *Journal of Computational and Applied Mathematics*, 185:292–307, 2006.

[110] J.H. Williamson. Low-storage Runge-Kutta schemes. *Journal of Computational Physics*, 35:48–56, 1980.

[111] Y. Xia, Y. Xu and C.-W. Shu. Efficient time discretization for local discontinuous Galerkin methods. *Discrete and Continuous Dynamical Systems - Series B*, 8:677–693, 2007.

[112] Y. Xing, X. Zhang and C.-W. Shu. Positivity preserving high order well balanced discontinuous Galerkin methods for the shallow water equations. *Advances in Water Resources*, to appear.

[113] X. Zhang and C.-W. Shu. A genuinely high order total variation diminishing scheme for one-dimensional scalar conservation laws. *SIAM Journal on Numerical Analysis*, 48:772–795, 2010.

[114] X. Zhang and C.-W. Shu. On maximum-principle-satisfying high order schemes for scalar conservation laws. *Journal of Computational Physics*, 229:3091–3120, 2010.

[115] X. Zhang and C.-W. Shu. On positivity preserving high order discontinuous Galerkin schemes for compressible Euler equations on rectangular meshes. *Journal of Computational Physics*, to appear.

[116] X. Zhang, Y. Xia and C.-W. Shu. Maximum-principle-satisfying and positivity-preserving high order discontinuous Galerkin schemes for conservation laws on triangular meshes. *Journal of Scientific Computing*, submitted.

Index